T0189609

Zoos

Also by Keekok Lee:

PHILOSOPHY AND REVOLUTIONS IN GENETICS

Zoos

A Philosophical Tour

Keekok Lee

First published 2005 by
PALGRAVE MACMILLAN
Houndmills, Basingstoke, Hampshire RG21 6XS and
175 Fifth Avenue, New York, N.Y. 10010
Companies and representatives throughout the world

PALGRAVE MACMILLAN is the global academic imprint of the Palgrave Macmillan division of St. Martin's Press, LLC and of Palgrave Macmillan Ltd. Macmillan® is a registered trademark in the United States, United Kingdom and other countries. Palgrave is a registered trademark in the European Union and other countries.

ISBN 978-1-349-54071-6   ISBN 978-0-230-50380-9 (eBook)
DOI 10.1007/978-0-230-50380-9

This book is printed on paper suitable for recycling and made from fully managed and sustained forest sources.

A catalogue record for this book is available from the British Library.

Library of Congress Cataloging-in-Publication Data
Lee, Keekok, 1938-
    Zoos : a philosophical tour / Keekok Lee.
    p. cm.
    Includes bibliographical references.
    ISBN 978-0-333-71110-1 (hardback)
    1. Zoos—Philosophy. I. Title.

QL76.L44 2005
590.73—dc22   2005051356

10   9   8   7   6   5   4   3   2   1
14   13   12   11   10   09   08   07   06   05

Transferred to Digital Printing in 2009

Contents

Acknowledgements

I wish to thank all those kind people who gave me technical assistance in producing the typescript. Most of all, I wish to thank Daniel Bunyard, the philosophy commissioning editor of Palgrave for his extraordinary speed and efficiency in enabling the book to see the light of day.

Introduction

This book has an unusual take on zoos. It is a philosophical exploration of the concept of zoos, not, however, from the usual ethical angle of either animal welfare or animal rights, but from the ontological standpoint.[1] It demonstrates that the animals kept in zoos are, indeed, unique to zoos – they are not wild, nor are they domesticants in the classical understanding of domestication. Ontologically speaking, they are not tokens of wild species but of what may be called artefactual species. Such a provocative thesis also has controversial and radical policy implications for zoos, as it challenges those policies advocated by the World Zoo Conservation Strategy (1993) and the European Union Zoos Directive (1999). The book is written not solely for philosophers and philosophy students but in a manner which makes it readily accessible to zoo theorists and managers world-wide, as well as to anyone who has a professional or lay interest in zoos and their futures.

The central contention of this book is that the ontological status of animals in zoos is different from that of animals in the wild. The former are not wild, amounting to what may be called biotic artefacts, and constitute the *ontological foil* to the latter.

This claim is inspired by two sources: first, what may be called the zoological conception of animals; second, the commonly accepted official definition of zoos as collections of animal exhibits open to the public. The respective implications of each of these two accounts are then teased out. Such an approach leads, as already mentioned, to some surprising conclusions and to some interesting implications for policy consideration in zoo management. The ontological stance and arguments presented in this book are, in principle, not against zoos; zoos are acceptable provided they are prepared to admit that their exhibits are not wild but tame or immurated animals, that the individual immurated animal is not a token of

1

any wild species.[2] However, to admit this logic would entail that the only sound theoretical justification of zoos lies in recreation, and not in the more high-minded mission of education-for-conservation of wild animals, or of *ex situ* conservation, tasks which zoos necessarily cannot accomplish, as an ontological dissonance exists between, on the one hand, the immurated animals and their behaviour on view as exhibits, and on the other, the mistaken belief, on the part of zoos, that by looking at such exhibits, visitors would actually be learning about wild animals, their behaviour in the wild and the need to save them and their habitats in the wild. Ironically, it transpires that the zoo-visiting public may have a better intuitive grasp of the ontological issues at stake, as they seem to accept happily and simply the fact that they derive pleasure and enjoyment from looking at, indeed, a unique kind of animal – only to be found in zoos – animals which may look wild but are not wild, and which necessarily do not behave like those in the wild. For them, the zoo experience is indeed unique, but not, however, in the way that zoo theorists and professionals have in mind.

Chapter 1 sets the scene for the ontological exploration of zoos by distinguishing between five different conceptions of animals, namely, the lay person's conception, the conceptions presupposed respectively by the philosophy of animal welfare (Singer) and animal rights (Regan), the zoo's conception and the zoological conception.

Chapter 2 elaborates on the zoological conception which rests primarily on the Darwinian theory of natural evolution and its mechanism of natural selection, in the context of other equally relevant sciences, such as genetics as well as ethology and ecology. At the same time, it introduces certain key philosophical notions such as trajectory, the distinction between existing 'by themselves' and 'for themselves'. The scientific and the philosophical components combine to lay down a delineation of the ontological status of animals in the wild, which in the chapters to follow, is argued as constituting the ontological foil to zoo animals.

Chapter 3 begins the delineation of the ontological status of animals under captivity in zoos by disentangling the conceptual confusions in terms such as 'wild animals in captivity', a phrase which occurs and recurs in zoo literature. However, it is an oxymoron, as the animals under zoo captivity are *ex hypothesi* tame; 'tame' is the antonym of 'wild', in a crucial sense of the term.

Chapters 4, 5 and 6 begin to go beyond the mere conceptual contradiction to demonstrate in earnest that zoo animals are, from the ontological perspective, very different beings from animals in the wild. Chapter 4 takes seriously the definition of zoos as 'collections of animal exhibits open to the public', teasing out its logical and ontological implications; it also examines the ideas that zoo animals are necessarily exotic (in the technical sense) animals, and that the environment provided for them is also, therefore, necessarily 'exotic' in the sense that, at best, their exhibit enclosures are simulated, naturalistic environments which bear no resemblance to habitats in the wild. Chapters 5 and 6 go beyond the simulation of naturalistic environments to examine several crucial aspects of animal life under zoo management which bear no resemblance whatsoever to the lives led by animals in the wild: (a) the miniaturisation of simulated space (in some extreme cases, enclosure space is 10 000 times smaller than the home range of the animal in the wild); (b) 'hotelification' (a word coined to refer to the fact that food is not food as in the wild, which is hunted or foraged, but are substitutes prepared in zoo kitchens, served up in much the same way as one would put out food for one's pet); (c) medication to prevent suffering from discomfort or disease, but also to ward off death and prolong the lives of the animals under zoo management. All these features add up to the suspension of the mechanism of natural selection within the context of natural evolution.[3]

Chapters 7 and 8 pursue the implications of the suspension of natural evolution and natural selection in the context of zoo management. Chapter 7 argues that zoo animals are immurated animals – 'immuration' is a term coined to embody the key idea that they are, indeed, domesticated animals, even though it is true that they do not readily fall within the traditional, classical definition of domestication, and that there is a need to distinguish between two kinds of domestication. Chapter 8 reinforces this logic, arguing that immurated animals are biotic artefacts, by laying bare the ontological features of artefacts in general, and showing that the term 'biotic artefacts' is neither conceptually confused nor incoherent. It formally introduces the type/token distinction and shows that an individual immurated animal is not a token of a wild species, but of an immurated species. In other words, zoo animals are uniquely different from animals in the wild as well as from traditional domesticants such as cats and dogs.

Chapters 9 and 10 begin the task of teasing out the policy implications of the ontological thesis established in the preceding chapters by critically examining the justifications for zoos. The three so-called serious justifications – research, conservation (*ex situ*) and education – are shown to be defective and, indeed, even deeply flawed given the ontological dissonance between zoo animals and wild animals in the wild. Curiously, the so-called frivolous justification in terms of recreation seems to have survived critical scrutiny best of all.

Chapter 11 and the Conclusion draw the book to a close by reconsidering the five conceptions of animals distinguished in Chapter 1 in the light of the ontological insights yielded by the subsequent chapters. The single most important conclusion, from the policy point of view, which this exploration has come up with is that zoos do, emphatically, have a future, indeed, a thriving future but, ironically, not perhaps the future which the World Zoo Conservation Strategy (1993) and the European Union Zoos Directive (1999) envisage and strenuously advocate – zoos are neither the modern-day scientific Noah's Ark nor are they the Open University for education-in-conservation, as they are human cultural spaces which necessarily humanise the animals under their care and control. Their animal exhibits, in reality, offer clean, wholesome family fun and entertainment. That is why, world-wide, they are big draws and big business. Ironically, it follows that although zoos (via *ex situ* conservation) may not be truly relevant to the project of saving extant threatened wild species from extinction, they, unwittingly, play a role in adding to biodiversity – though not of the natural kind – by nurturing and creating, in the long run, new immurated, artefactual species.

1
What Does the Public Find in Zoos?

The question in the title of this chapter seems silly and naïve. However, there is a point in posing it, as behind the obvious answer to it stands a whole lot of complexities which must be unravelled before one could ultimately answer in a satisfactory manner the issues raised in the Introduction.[1]

Surely, even a child would be able to say that in zoos one finds animals. The child would be right in one undoubted sense. While there are trees in zoos, one does not go to a zoo specially to see plants; plants are incidental to the *raison d'être* and aims of zoos. Zoos exist to exhibit animals.[2]

However, the intriguing question is: 'What animals do zoos exhibit? All animals?' To which we may answer: 'Not at all; only some select few.' In other words, the selection presupposes a certain conception of what counts as an animal. We need to explore that conception, and compare it with at least three other conceptions, bringing out the overlapping concerns and relationships (if any) between them, making clear their respective hidden agendas and assumptions, whenever relevant.

The lay person's conception

Let us start with the ordinary lay person's conception of what counts as an animal. The first thing to notice is that it sticks to the commonsensical understanding of the traditional two-kingdom schema (as evidenced in the obvious answer to the naïve question mentioned above); they would have no difficulty classifying squirrels as animals and conifers as plants, but on the whole they would have no opinion whether bacteria count as animals, since they would have no knowledge nor would they have given the matter any thought.

5

In general, society's interest in certain animals was/is dictated by the roles they play in human lives – these animals have either religious/cultural, culinary, economic or personal significance for the social group or individual in question. For instance, some groups have chosen even rats (vertebrate) and snakes (invertebrate) as objects of religious worship. Some cherish the bald eagle as a symbol of national (tribal) virility, others the lion. Which animals are good to eat and which are not clearly varies according to culture and to historical period. Dogs are good to eat for the Dayaks (in Borneo), while cows, for Hindus, are not for eating at all. Tigers, rhinoceroses and some whales, today, are in danger of being hunted to extinction for economic reasons.

Beyond identifying animals which belong to these main categories of concern, most people remain in ignorance of those not encompassed. Today, in the industrialised and industrialising world, 'animals' as a generality are not all that pertinent to their lives. Particular types of animals may, outside these categories, catch their attention because they are exotic (in which case they go to the zoo to see them or watch them in a television programme), charismatic like the lion, or cuddly like the panda. As far as lay people are concerned, birds seem to be the only class of animals which commands a sizeable minority of followers dedicated to watching and studying them; and amongst ornithologists knowledge of them can be thorough and comprehensive.

As most societies have long left the hunter/gatherer mode of existence behind, the only animals which form part of their immediate experience and consciousness are domesticated animals – cows, pigs, sheep, goats, chickens (which are good to eat); horses, bullocks, camels (which are good for traction and transportation but also to eat in some cultures); and dogs and cats (which, in the West, are good for companionship, guarding the house or catching mice but not good to eat). Except for chickens, ducks and turkeys which are birds, the rest are mammals. In other words, the word 'animal' would, in the context of utility derived from domestication, typically conjure up either of these two classes of the phylum chordate. In some cultures, some species of fish have been domesticated, and increasingly today salmon and trout are also being cultivated. However, domestication, throughout world history, has been confined to only a few species of these vertebrate classes.

In contemporary consciousness, the image or idea of what counts as an animal has become even more circumscribed, as increasingly in urban contexts, domesticated animals are not directly encountered. Some children even have difficulty associating pork with an animal called the pig or milk with the cow, pork and milk being just packaged items the family purchases from the supermarket. This means that animals as domestic pets occupy centre stage; especially in developed countries, cats and dogs are the most prevalent. Children, increasingly, are taught to identify animals via these exemplars. For them, the denotation as well as the connotation of the term 'animal' is paradigmatically given and determined by the varieties of dogs and cats they find in the household. If asked whether they share their homes with animals, they would confidently say no provided they kept no cats, dogs, budgerigars, goldfish or hamsters. If reminded that most homes, and therefore, theirs, would have a mouse or two, they would feel justifiably shocked. But to them, if compelled to acknowledge their presence, mice are not animals in the way pets are animals – they are at best animals only in some technical sense. To them they are just pests. And if told that mites live in the detritus of their scalp or upon their skin, and in their carpets, they would be horrified; unlike mice or rats, they would even have difficulty accepting or understanding these as animals at all.

To sum up, the lay consciousness increasingly is confined to grasping animals in terms of a few domesticated species of mammals which are regarded as friends to humans or of a few exotic and/or charismatic animals which they see occasionally in zoos (as we shall see).

Conceptions presupposed by animal welfare/liberation and animal rights

We next turn to the conception of what counts as an animal via the defence of animals against human cruelty; it is essentially a protest against the ways in which animals are: (a) kept/caught and then slaughtered for food or other human purposes; (b) used in scientific research and experimentation whether for the serious purpose of saving human lives or the relatively more trivial one of improving the appearance of human bodies; and (c) hunted, hounded or killed for human pleasure.

The more traditional justification, derived from philosophers like Kant, is that the duty not to be cruel to animals is in reality an indirect duty to humans, as the infliction of cruelty upon animals could dispose us to be callous towards fellow human beings. But of late, this highly anthropocentric standpoint has been powerfully challenged by two contemporary philosophers – Peter Singer (1976) and Tom Regan (1983) – who, in spite of the obviously different philosophical stances each has adopted, nevertheless are united in repudiating the dominant humanist tradition of Kant and the Enlightenment, at least regarding the treatment of animals.

A minimalist reconstruction of Singer's philosophy of animal welfare includes the following:

(a) *The hedonic postulate.* Pleasure and pain as mental states are respectively intrinsically good and evil.
(b) *The consequentialist/utilitarian postulate.* One ought always to maximise pleasure and minimise pain in one's actions.
(c) *The boundaries of sentience postulate.* (a) and (b) are 'blind' to the kind of being which is capable of feeling pleasure and pain. Humans clearly are sentient but empirically it can be shown that humans are not the only sentient beings. Other mammals, too, clearly are sentient. Birds are as well. Erring on the side of caution and charity, the boundary should then be drawn somewhere around shrimps and, possibly, lobsters.
(d) *The consistency postulate.* As we, today, believe that we have a moral duty not to keep and eat fellow humans for food, to perform vivisection on them with or without their consent, to hunt, maim or slaughter fellow humans for entertainment, then equally, we have a moral duty not to do likewise to fellow sentient beings.

A minimalist reconstruction of Regan's philosophy of animal rights includes the following:

(a) *The rights postulate.* (i) an entity is intrinsically (or inherently, in Regan's terminology) valuable if and only if it is capable of being the subject of a life, that is to say, possessing memory, beliefs and desires as well as other mental states, and (ii) an entity is a rights holder if and only if the entity is capable of being the subject of a life.
(b) *The conceptual postulate.* To be the subject of a life, to experience mental states like beliefs and desires, conceptually speaking, it is not necessary to possess verbal language at all or human language as we understand it to be.
(c) *The boundaries of the subject-of-a-life postulate.* (a) and (b) are 'blind' to the kind of entity which can satisfy the criterion of being the subject of a life. Humans (or at least the majority of them) are clear candidates, but empirically it can be shown (once (b) has been conceded) that mammals, too, are candidates. Erring on the side of caution and charity, the boundary of eligibility should then be drawn at birds.
(d) *The consistency postulate.* As we, today, hold the view that human beings have a right not to be kept and eaten by fellow humans, to have vivisection performed on them with or without their consent, to

be hunted, maimed or slaughtered by fellow humans for entertainment, then equally, other mammals (and possibly birds) have a right not to be treated likewise by us humans.

There are obvious differences in the philosophical foundations provided by Singer and Regan and the debate which ensues between the two sides – one is anchored in moral duties understood in the context of hedonic consequentialism, the other in moral rights, deontologically understood, in the context of certain characteristics of mental life in humans and closely related mammalian others.[3] However, these differences notwithstanding, the two do have certain things in common, apart from their agreed common goal to end cruelty to, and the suffering of, animals. Their respective implicit conceptions of what is an animal are given by the criterion they each have chosen as the most fundamental postulate of their philosophy of animal liberation – the hedonic postulate in the case of Singer and the rights postulate in the case of Regan.

In either, the paradigmatic animal is the human animal. Although Bentham, as the commonly acknowledged founding father of modern utilitarianism, had said that certain animals also come within the purview of his fundamental postulate, nevertheless, utilitarianism, as propagated and inspired by him, has chosen to concentrate on humans as the paradigmatic sentient beings. Similarly, the concept of rights – understood either as natural or contractual rights – has long been conducted, until very recently, within an exclusively human domain.

Singer uses the image of the expanding (moral) circle, in order to draw certain other beings, so far excluded by modern Western philosophy, into its orbit. Regan endorses this implicitly. However, both proceed on the assumption that there is a limit to which this circle may be enlarged – Singer's fundamental postulate allows him to redraw it with a somewhat wider radius than Regan's. But in the centre of their circles is the human. The further a being is from that centre, the more difficult it would be to make a case for bestowing on it the status of being morally considerable.[4] The human is, of course, a mammal. Hence, extending moral duties or rights to fellow mammals is their most obvious target. This has prompted some commentators to say, especially in the case of Regan's account, that it is really about mammalian rights.

In general, it might not be too unfair to say that both philosophies are underpinned by an overarching postulate, namely the search for similarities and likenesses between humans and certain animal others.[5] As such, the more an animal resembles humans in certain specified ways, the easier it is to argue for admitting them into the moral circle.

Of the mammals, the Great Apes come closest to us – this class is held to consist of the gorillas, the orang-utans, the chimpanzees and then ourselves as the long-missing fourth Great Ape.

While those animals within the pale are accorded a dignity befitting their newly acquired status of being morally considerable, those outside, as a result, are dealt a double blow – first, they are owed no moral duties and denied moral rights, and second, the term 'animal liberation' or 'animal rights' itself goes even further and serves implicitly to deny them the status of animality itself. In other words, only those beings which qualify to be the bearers of rights or to be the object of our moral duties are 'proper' or 'true' animals. The denotation and connotation of the word 'animal' has surreptitiously and subtly been revised so that even on Singer's more hospitable expansion of the moral circle, worms, molluscs and many more are debarred. The similarities postulate has forcefully challenged human chauvinism, the view which sets humans apart from other animals, assigning to themselves a superior status of privilege and domination.[6] It attempts to force human consciousness to concede that humans, as mammals, are really fellow animals. They (together with those admitted into the expanded circle) are owed duties not to be tortured and held to enjoy rights to life, and so on. Strictly speaking, in Singer's moral/political philosophy, a single hedonic consequentialist theory is postulated to embrace all sentient beings, from mammals down the evolutionary scale to possibly some crustaceans like lobsters, just as in Regan's moral/political philosophy, a single unified theory of rights is postulated, covering all mammals and possibly birds. However, the price for this revision is the construction of a new demarcation line between the ingroup and the out-group. Members of the latter are pariahs because they are unlike us in crucial respects, and therefore cannot be animals, a category to which we, ourselves, now belong. A hierarchical or class system remains in place – the franchised and the privileged against the non-franchised and the disadvantaged. It is just that the former now includes not simply us, but those beings which are like us in certain selected aspects. Human chauvinism may have been vanquished but its spirit has not been challenged by either Singer or Regan and remains unexorcised in their respective philosophies.

The zoo's conception

We have seen how the denotation and connotation of the term 'animal' endorsed by the lay conception as well as that of Singer or Regan permit the lay public as well as animal theorists such as Singer and Regan to be selective in what they count as animals. The same selectivity holds in the case of the zoo conception of animals. However, it differs from the lay conception in that zoos largely exclude domesticated animals, while the lay public includes them in their account; otherwise, it concurs with it in all other respects.[7] In the words of one prominent apologist for zoos, zoos deal in the main with animals like 'the rhinos, the tiger, leopards, primates, parrots, the Asian elephants, many an antelope, many a bird of prey, various cranes and so on; all the creatures of our childhood; what most people mean by the word "animal" ' (Tudge, 1992, p. 46).[8] However, it may be fair to say that while both the lay public and zoos consider zoo animals to be exotic, for the former, the term 'exotic' carries the connotation of being from distant parts, uncommon, strange, or rare, whereas for the latter, the term is understood in the more technical sense of referring to the geographical origins of the animals, to the fact that their natural range is found in parts of the world other than where the zoo and its exhibits are.

The zoo image, then, powerfully reinforces the lay image of what counts as an animal. What zoos and ordinary people mean by the word 'animal' and the species the animals belong to, refers, then, only to a minuscule fraction of the total animal species known today to science; that number stands at 1 032 000 species, of which insects account for nearly three-quarters, standing at 751 000 species. Mammalian species only total 4000, with birds slightly more than double that at 8600.

The zoological conception

To get a more comprehensive conception of what counts as an animal we need to turn to zoology, which is commonly understood as the scientific study of animals. One of the Greek words composing the term itself – *zoon* – is usually translated to mean 'animal', although it may be more accurate to point out that it has a wider denotation, referring to living things.[9] Of course, zoology in turn is part of biology, the study of life itself – the Greek word *bios* means life.

So how does zoology answer the question 'what counts as an animal?' It will soon be obvious that as far as it is concerned, there is no simple and quick reply. Any systematic answer, no matter how schematic, starts with the, by no means easy, problem of first distinguishing between life and non-life. Although both the living and the non-living are subject to the same laws of physics and chemistry as well as the law of the conservation of energy, the crucial differences between them lie in the fact that the former is very differently organised and structured from the latter – unlike the latter, it is capable of metabolism, growth, adaptability, irritability and interaction with the environment.

But how in turn is animal life to be distinguished from plant life? As the terms 'zoology' and 'botany' themselves indicate, we, lay people, take for granted that there are two recognisable kingdoms to which all organisms are said to belong: plant or animal. We instinctively know to classify mosses, ferns and trees as plants on the one hand and mammals, birds and fishes as animals on the other. Yet this time-honoured Aristotelian schema may be said to have outlived its usefulness in the light of more up-to-date understanding of the various forms of life on Earth. Complexities appear straightaway. The central point to grasp is that, unfortunately, no single criterion exists which can serve to distinguish all animals from all plants. Take the presence or absence of chlorophyll as an obvious distinguishing mark. Chlorophyll is a necessary condition for photosynthesis to take place. We unhesitatingly associate chlorophyll with plants but not with animals; an oak has it but not a hedgehog. Under photosynthesis, green plants produce organic compounds from sunlight and atmospheric carbon dioxide and, at the same time, restore free energy to the biosphere. These photo-autotrophic organisms, in converting inorganic substances into organic materials, not only sustain their own functioning integrity but also provide food for heterotrophic organisms, namely animals, which live on them as they themselves lack the capability of photosynthesis. Yet some organisms, for example *Euglena*, display photosynthesis under some conditions but not others – in the light, it functions like a plant, in the dark, like an animal. So is it an animal or a plant? They are considered to be animals by zoologists and plants by phycologists. Another borderline group is the slime moulds – zoologists call them Mycetazoa and botanists Myxomycophyta. Furthermore, not all plants possess chlorophyll – the higher parasitic plants and a large plant group, Fungi, also do not have it. So the presence of chlorophyll cannot identify and include all plants; neither does its absence identify all animals.

Another distinguishing mark may be said to be motility. We common-sensically believe that animals have the ability to move about in their environment (and some even travel between very different environments depending on the season) at some stage in their life history, whereas plants are stationary. Yet movement is not restricted solely to animals – a good many of the thallophytes such as Oscillatoria, several bacteria and colonial chlorophytes are quite motile.

In biology today, scientists, as we have seen, no longer regard the two kingdom schema to be all that illuminating; instead, they attach greater significance to the prokaryote/eukaryote distinction. Prokaryotes refer to organisms which lack organelles, that is, specialised structures such as nuclei, mitochondria, chloroplasts; their DNA is not found in chromosomes but forms a coiled substance called a nucleoid. In contrast, the genetic material of eukaryotes is contained within a well-defined cell nucleus with a protein coat. However, besides this crucial difference, there are others. All living organisms, except viruses, bacteria and blue-green algae (cyanobacteria) are eukaryotic.

The problems mentioned above, amongst others, led to a proposal in 1969 for a five-kingdom system which incorporates the prokaryote/eukaryote distinction. The prokaryotes are assigned to the kingdom Monera while the eukaryotes are divided into four kingdoms. The kingdom Protist includes the unicellular eukaryotic organisms (protozoa and unicellular eukaryotic algae). The multi-cellular eukaryotic organisms are in turn organised into three kingdoms according to their mode of nutrition and other significant organisation differences. The kingdom Plantae contains multi-cellular photosynthesising organisms, higher plants and multi-cellular algae. The kingdom Fungi includes yeast, moulds and fungi that get their food through absorption. Finally, the kingdom Animalia comprises the invertebrates (except the protozoa) and the vertebrates.

Traditionally, the animal kingdom has included the unicellular protozoa, but the new schema excludes them. Yet they share many characteristics with so-called animals, such as ingestion of food, advanced locomotory systems, sexual reproduction, and so on. For this reason, books on zoology regard the protozoa as animals and have a chapter on them.

To sum up a very complex set of issues, it may be fair to say that zoologists today clearly identify all invertebrates and vertebrates as animals while agreeing that the protozoa, too, may be considered as such.

One should also be reminded that the consensus which emerges takes place against a background of theoretical ideas and assumptions which have developed since the nineteenth century, of which the most salient are: (a) the Darwinian theory of natural evolution in terms of natural selection; (b) the Mendelian theory of particulate inheritance and the gene/chromosome theory; (c) DNA genetics and molecular biology;[10] (d) developments in cell theory; (e) animal ecology; (f) ethology; and (g) an understanding of intra-ecosystemic interdependence.

The larger framework is still basically neo-Darwinian. It is informed by the notion of natural selection as the mechanism of natural evolution. Its primary object of study is therefore animals in the wild, to understand their ancestry and evolutionary history, how they come to have the characteristics they posses through certain fundamental concepts and principles that govern the understanding of organic life in general and animal life in particular.[11] This basic orientation is informed by four deep themes in the philosophy of the (wild) organism:

1. *Biology of time.*   Evolution cannot be detached from time and history. As Steven Rose has put it: 'Nothing in biology makes sense except in the light of *history*, by which I mean simultaneously the history of life on Earth – evolution, Darwin's concern – and the history of the individual organism – its development, from conception to death' (Rose, 1997, 15).

2. *Biology of space.*   Added to the time dimension is the space dimension, which in organisms is expressed through their structures. Organisms have forms which undeniably change over time, yet they persist as they go through their trajectories in life.[12]

3. *Biology and philosophy of reciprocal causation of organisms-in-the-environment.*   Organisms cannot be considered apart from their environment.[13] Organisms are the product of a complex set of dynamic causal interrelations between themselves and their environments. As Rose has put the matter so well, it is appropriate to quote him:

> for the organism, (the environment) is the external physical, living and social worlds. Which features of the external world constitute 'the environment' differ from species to species; every organism thus has an environment tailored to its needs. . . . *organisms evolve to fit their environments, and environments evolve to fit the organisms that inhabit them.* No environment is constant over time. . . . Stasis is death. . . . Organisms – any organism, even the seemingly simplest – and the environment – all relevant aspects of it – interpenetrate. Abstracting an organism from its environment, ignoring this dialectic of interpenetration, is a reductionist step which methodology may demand but which will always mislead. . . . organisms are not passive responders to their environments. They actively choose to change them, and work to that end.
>
> (Rose, 1997, p. 140 [my emphasis])

It is not good science to regard an animal in abstraction from its evolutionary history and context or from its natural habitat which informs its very existence and with which it interacts.[14]

4. *Environmental philosophy of ecocentrism.*   This philosophy focuses on the conservation of (natural) biodiversity, on the survival of animal

species rather than of individual animals, irrespective of whether or not they are charismatic, exotic, capable of suffering pain or of mental activity.[15]

The remainder of the book will tease out the implications of the zoological conception of animal in order to throw ontological light on the respective status of wild animals in the wild on the one hand, and captive animals in zoos on the other.

Conclusion

This chapter has briefly set out five conceptions of what counts or does not count as an animal. We shall have occasion, as the arguments in the book develop, to return to them in order ultimately to assess the different justifications for zoos; we shall also have to go beyond this initial exploration of the different denotations and connotations of the word 'animal' as embodied in these conceptions to unravel in greater detail the conception of zoo animals as opposed to that of wild animals (not in captivity), before attempting to answer the questions and address the issues raised in the Introduction.

2
Animals in the Wild

This book in the main is an attempt to make explicit the presuppositions and tease out the implications of the zoological conception of what counts as an animal, in order to clarify the ontological difference between wild animals on the one hand and zoo animals on the other. This chapter begins the task by looking in greater detail at animals in the wild in order to characterise their ontological status before turning attention in the chapters that follow to determining in turn the ontological status of animals in zoos; it is a fundamental contention of this book that the former is the ontological foil to the latter, a point which will become clear as the arguments unfold.

Natural evolution and natural selection

The zoological conception so far outlined of what counts as an animal has made it clear that animals in the wild are the products of natural evolution. Natural evolution means that all forms of life as we know them to be today and historically are/were descendants of the first form of life on Earth. Once upon a time, nearly 4 billion years ago, Earth was more or less devoid of life. When life did appear, it was first in water as microbial mats. The first organism was prokaryotic and single-celled. Then the 'higher' eukaryotic organisms appeared about 1.8 billion years ago, at first as single-celled, later as multi-cellular. It was not until the Cambrian explosion, 540 to 500 million years ago, that macroscopic animals appeared in abundance to give rise to the types which still exist today. Apart from the protozoans, as already observed in Chapter 1, the kingdom Animalia comprises the vertebrates and the invertebrates, dating largely from the Cambrian period.[1]

According to natural evolution, the beginning of all life forms and, in principle, the end of all life forms on Earth have nothing to do with either gods (God) or humans. No divine agency of any kind is necessary to account for the origin of life and its evolutionary history.[2] It relies on natural selection to account for the latter. Natural selection – that mechanism working upon genetic variations in individual organisms which constitute populations of organisms interacting with other components of their ecosystems – primarily explains the course of evolution.[3] Their origins pre-dated human existence by about 4 billion years and their evolution, until very recently, also occurred unhindered or relatively unhindered by human activities. However, the human species today, in virtue of its success as a species with a population of over 6 billion and still growing, and its present unconscionably extravagant mode of production and consumption in terms of exhausting Earth's resources, especially amongst the world's mature economies, is feared to be capable of destroying most of Earth's (natural) biodiversity. Under the more pessimistic scenarios, it is even capable of almost wiping out the kingdoms of Plantae and Animalia (including the human species) should some crazy state in possession of nuclear bombs release them in the spirit of *après moi, le déluge*.[4]

Trajectory and the distinction between 'by themselves' and 'for themselves'

The brief account of natural evolution and natural selection given above must be supplemented by an exploration of a new notion, namely trajectory and the distinction, inherent in it, between beings which exist 'by themselves' and beings which 'exist for themselves'.[5]

Every naturally occurring entity or process has its own trajectory. The term 'trajectory' is introduced to do the precise job of referring to the history as well as the character of its autonomous existence, whatever the entity or process may be. For example, a lake has its own trajectory. As a geological form, it is considered to be one of the most transient – it dries out in a relatively short span of geological time, first by becoming a swamp, then probably a meadow. In the case of naturally occurring processes, these may be biotic, abiotic, or an interaction involving both. It is also the case that a species (for instance, *Canis lupus*), as much as an individual member of the species, (the particular wolf roaming at a particular time in a particular forest), may be said to have their own respective trajectories, although as we shall see in the next section, the trajectory of the species is not necessarily identical to that of any of its individual members.

Similarly, the use of the term to cover both the biotic and abiotic domains does not entail that their trajectories are identical, that what is true of the one is also true of the other. We have already seen that, as far as present evidence goes, the history of Earth shows that the abiotic long preceded the biotic on Earth. Present evidence also shows that without the continuing existence of a certain combination of abiotic conditions, the biotic would not, and could not, continue to exist.

Two further points need to be emphasised. One concerns the crucial differences between the biotic and the abiotic; the other is to argue that in spite of these differences, it is appropriate to use the notion of trajectory to talk about the entities and the processes in both domains.

First, the crucial differences: in general, individual organisms go through certain recognisable stages from their beginning to their end – infancy, growth, maturity, senescence (in some cases) and death. They also possess certain characteristics, depending on the particular stage of their existence or indeed, of their sex. For instance, in the trajectory of a frog, it starts off life by being an embryo, which soon develops into a tadpole, then an adult frog. The tadpole clearly looks very different from the adult frog; yet the former is but a stage in the growth of the latter. Similarly, the peahen looks very different from the peacock; yet both are members of the same species.

By contrast, the granite mountain remains a mountain of granite. Of course, even granite wears away over a large expanse of geological time. But granite does not 'mature' to become something else, although it is true even granite might weather away to become soil. However, it does not 'grow or develop to become soil' in the same way as the tadpole becomes the adult frog – hence the need, strictly speaking, to put quotation marks around the phrase in the former but not in the latter context. Granite may end up as soil, but soil is not granite – they are two very different things. However, in the case of the frog, the tadpole and the adult frog belong to the same species; each is simply a different stage in the trajectory of the individuals belonging to the species.

Another obviously important difference between the biotic and abiotic is that the former appears to be an exception to the laws of thermodynamics, whereas the latter is not. But, of course, the appearance is only misleading. However, what misleads one is that there are certain processes at work in organisms, which are absent in the case of the abiotic. Individual organisms are autopoietic; they sustain and maintain their own functioning integrity by engaging in metabolical and other physiological activities. They, therefore, appear to produce order out of chaos, so to speak; in reality, to produce order, they need to take in ordered material from external sources, in the form of nutrients which, ultimately, are dependent on solar energy itself. They produce disorder when they respire, defecate and when they die. In contrast, abiotic entities are not autopoietic or capable of self reproduction, as they do not possess any mechanisms analogous to those found in organisms.[6]

However, in spite of the admitted differences, one would want to argue that the term 'trajectory' may meaningfully be applied to both. Only the biotic may be said to be autopoietic, yet both biotic and abiotic nature may be said to be 'self-sustaining' and 'self-generating' in the larger sense of these terms, in spite of the fact that the latter does not possess physiological/metabolic, neurological, hormonal or reproductive mechanisms which the former has. In other words, the autopoietic is but a subcategory of the self-sustaining and the self-generating.

The abiotic as much as the biotic is 'self-sustaining' and 'self-generating' in the sense that it is autonomous, though not autopoeitic.[7] But, to be more precise about the autonomy of the different trajectories in general of biotic and abiotic nature, one needs to explore an important distinction between saying that an entity exists 'by itself' on the one hand, and that it exists 'for itself' on the other. While all naturally occurring entities, whether biotic or abiotic, exist 'by themselves', only the former exist 'for themselves'.

Naturally occurring entities and processes are precisely those which have come into existence, continue to exist, and go out of existence, entirely autonomously, and therefore independently of human intentionality and agency (and of supernatural agency for that matter). They do not owe their being in any way to humankind. They are also self-generating and self-sustaining. So, in so far as they exist, they can then be said to exist 'by themselves'. However, naturally occurring biotic entities display an additional dimension of complexity in their existence. In existing 'by themselves', they also necessarily exist 'for themselves' given that they are members of organic species, unlike abiotic entities which are members of inorganic natural kinds.[8] As organisms, they strive to maintain their own functioning integrity. They develop, grow, mature, replicate (and in the case of the higher animals, they nurture their young) – their success in these activities involves active appropriation of suitable nutrients, protecting themselves against adverse circumstances which would either harm or kill them.

Implications of natural evolution/natural selection in the light of the notion of trajectory

It follows from the historical fact of natural evolution and the explanatory mechanism of natural selection which underpins it that animals in the wild are:

1. Naturally occurring entities: both as individual organisms and as species, they have come into existence, continue to exist, and will eventually go out of existence (in principle), entirely independent of human volition, manipulation and control.[9] In contrast are those entities which have come into existence, continue to exist and go out of existence precisely as the result of human intentions and manipulation. In the category of naturally occurring entities, the term 'nature' is understood in its fundamental sense as the ontological foil to the latter category which refers to human artefacts.[10] Human artefacts may be briefly defined as the material embodiment of human intentionality; as such, they are technological products, and may be either abiotic/exbiotic (like statues made of marble or wood), or biotic (like domesticated plants and animals).[11] In the case of biotic artefacts, the technology used is initially craft-based (as in the classical case of traditionally domesticated animals) or, latterly, science-induced (first based on the classical science of Mendelian genetics in the first half of the twentieth century and then on the sciences of molecular genetics and molecular biology over roughly the last 40 years).[12]

2. To use slightly different language, one could say that animals in the wild (both as individual organisms and as species), as characterised above, exist 'by themselves'. So do abiotic entities, such as mountains and rivers. However, unlike the latter, biotic beings, such as (individual) animals in the wild, may be said also to exist 'for themselves'. By this is meant that they maintain their own metabolism and functioning integrity in the total absence of human intervention and manipulation; they keep themselves healthy, keep disease at bay until the moment comes when they fall foul of disabilities, ailment and/or fall prey to predators because of such weaknesses. Being self-maintaining, self-regenerating and self-reproducing, they may also be said to be 'autopoietic' beings.[13]

3. In reproduction in general, the male fights off other contenders in order to mate with the female he fancies or encounters; the female, too, up to a point, chooses to mate with the male she fancies, although she usually ends up with the victorious male who has succeeded to fight off his rivals. Their reproduction is under their own control. Depending on the species, but generally speaking, the females are responsible for bringing up their young and initiating them into the way of life which is normal for the species to engage in. For example, the young will be taught to hunt if the species is carnivorous; to identify the right plants for foraging if the species is herbivorous; to learn about what constitutes danger; to get to know the terrain within its home range, and so on. Just to cite one example of such education: the mother orang-utan teaches her offspring, from its birth till the age of 7, a whole range of survival skills. The infant learns how to navigate the forest by learning how to make mental maps of distances, of food locations. The mother teaches it to identify and to obtain up to 400 different foods, as well as to swing from tree to tree with immense dexterity, to judge finely subtle variations in the strength as well as the suppleness of branches. It is a prolonged and intensive education, as if of an only child undertaken by the mother.[14]

4. The species determines the social composition and life of the group, whether it is a group predominantly of mother (and a few other females) and their young, excluding any mature male who normally leads either an isolated life of his own (except during the mating season when he seeks contact with females) or as member of a small group of other similar males.

5. In other words, every individual animal in the wild (as a member of a species) unfolds its own *telos*, its own trajectory from birth till death, as well as that of its species. It may be said to exemplify the thesis of intrinsic/immanent teleology.[15] The tiger hunts, while the giraffe forages; the male lion lives in a pride with a few other sexually mature males, while the emperor penguin shares the duties of parenthood equally.[16]

6. Biodiversity may be a recently coined term but the notion refers to something which is of fundamental concern to scientists interested primarily in evolution leading to speciation – for instance, a single species of wasps which came to Hawaii 100 000 years ago has given rise to hundreds of species as the members of the original colonising population spread out, changed and evolved in response to the distinctive environments they found themselves in, which were peculiar to a particular island, mountain ridge or valley. As such, the scientists are interested, not so much in the individual animal (or organism) but in the species

and in the mechanisms of speciation, namely how changes in a population of individual organisms lead eventually to the emergence of two or more populations which no longer exchange genetic material with one another.[17]

Evolution of species means that a population responds not merely to genetic variations but to such variations in the context of specific environments. Over time, variations which prove to be adaptive may ultimately lead to the emergence of two or more species, as we have seen. This means that ecology in general, and habitats and ecosystems in particular, play a vital part both in the emergence and the maintenance of a species – such a view, in environmental philosophy, is also referred to as holism or ecocentrism.

According to this scientific understanding of animals (in the wild), behind the individual animal necessarily stands the species as well as the habitat of which both the individual and the species are a part. What one observes of the individual animals cannot be properly comprehended except in relationship with their species and, in turn, of both in relationship with the environment in which they have their being and function. As the example of the wasps in Hawaii demonstrates, the populations in the wild which constitute different species have evolved in the particular habitats they had found themselves in and to which they had adapted. The complex interactions between the different environments and the individual/species historically had led to their different evolutionary trajectories.

For the same reason, one cannot understand what a polar bear is, and what the species *Ursus maritimus* to which the bear belongs is without having an idea of the Arctic landscape and seascape, and the rapid series of evolutionary changes which its ancestors underwent in order to survive the cold.[18] Scientists believe that the polar bear is a descendant of a group of brown bears and is the most recent of the eight bear species; this group was stranded and isolated by glaciers in Siberia during the mid-Pleistocene period, that is, 100 000 to 250 000 years ago. While brown bears hibernate in the winter, polar bears do not, strictly speaking, do so, as food is plentiful in the Arctic winter. Their bodies are more elongated, and their faces have different shapes; their necks are longer which enable them to keep their heads above the water while they swim; their teeth are different as they are carnivores whereas brown bears are, in the main, if not exclusively, herbivores; they have warm thick fur and very large paws which make it possible for them to spread their weight on thin ice as well as to facilitate swimming. A polar bear's home range can be enormous, its size being normally dependent on the amount of food available – in one case, it is twice the size of Iceland, and in another, one satellite-tracked female travelled 3000 miles. A young polar bear may travel a distance of as much as 1000 kms (600 miles) in order to set up a home range of its own when it leaves its mother, after an apprenticeship of two to three years during which its mother teaches it how to catch seals, to respond to the seasonal variations in the availability of food, and so on.[19]

7. The individual animal is but a very transient member of its species. A species, as Holmes Rolston (1988) puts it, is a historical lineage. It comes to possess the characteristics it does as the outcome of an extended period of evolution which, in the case of the polar bear, as we have seen, began 100 000 to 250 000 years ago. Hence an individual animal properly understood against the background of its species is not an *ahistorical* being, as it is the product and an embodiment of evolutionary history itself. In other words, in observing a particular animal, one is not merely observing an individual being displaying whatever characteristic it does possess, but through it one grasps the whole historical dimension of its evolutionary past. This understanding of species refers to the evolutionary-species concept.[20]

The individual animal, such as the polar bear, may be said to be a token of its type or species (both as it exists today and in its evolutionary history) *Ursus maritimus*. But what is the precise relationship in this biological context between a token and its type?[21] First of all, membership of an animal species does not demand that all tokens are absolutely homogenous; membership can and does tolerate variations between individual tokens.[22] However, the fact that one token weighs more than another, that it is taller or longer than another, or that it may even look very different from another does not prevent them from being tokens of the same species.[23] Second, certain attributes of tokens cannot intelligibly be used of their species. For instance, male polar bears may weigh around 600 kg while the females are much smaller, around 400 kg. In other words, we can catch and weigh individual polar bears and record their respective weights; however, it makes no sense for us to say that we can catch and weigh the species. Third, although it makes sense to assign both birth and death dates to tokens and their species, it is true to say that while one can (in principle, especially today with the help of telemetry) record the precise birth and death of a token, one cannot (in principle) determine with any degree of precision the emergence or the extinction of a species. Fourth, an individual elephant is likely to live up to 65 years but the life span of its species is more likely to be a million years.[24]

In spite of some of the obvious differences between tokens and their species noted above, all the same there is a tight existential relationship between them. First, if no tokens whatever exist, then the species could be said to be extinct. However, in contrast, if only a few tokens exist, the species may be said to be as good as extinct, especially if only a few exclusively male or exclusively female tokens remain in the case of a sexually reproductive species. Even if a small population of male and female tokens of a reproductive species are extant, biologists would still, nevertheless, consider the species as doomed to extinction, as the collective genetic inheritance between them is too limited to make the species viable in the long run. Second, it follows that existentially speaking, a species is not something over and above its tokens – there is no immortal, eternal species existing in some Platonic heaven apart from its transient mortal tokens. Third, the extant tokens instantiate, at once, their own identity and trajectory as well as the identity and trajectory of their species.

Conclusion

This chapter seeks to establish two main points:

1. It argues that to grasp what an animal in the wild is, one must understand not simply what constitutes its own identity (in terms of what it looks like, its full range of behaviour), its trajectory, but also its identity and trajectory in relation to its species within the habitat and ecosystems to which it reacts/reacted in a complex manner here and now, as well as in evolutionary history. In other words, an animal in the wild must be suitably *contextualised*. The individual polar bear is a polar bear not simply because it looks a certain way, has certain anatomical/metabolic characteristics, behaves in certain ways with regard to securing its livelihood, its reproductive activities, to looking after and initiating its young in survival skills, but also because it is a token of a species which has evolved some 100 000 to 250 000 years ago from its ancestor – the brown bear – in the Arctic habitat, landscape and seascape.

2. It demonstrates that from the *ontological* standpoint, animals in the wild are *naturally occurring* beings; they are not the products of human design, manipulation or control, as their very existence and survival have nothing (in principle) to do with humankind.

3
'Wild Animals in Captivity': Is This an Oxymoron?

The last chapter attempted to shed light on the ontological status of animals in the wild as naturally occurring beings. The next five chapters will try analogously to clarify the ontological status of zoo animals. However, this chapter will grapple specifically with a basic problem of terminology, whether it is conceptually appropriate to refer to animals in zoos as 'wild animals in captivity'. It will argue that the term is an oxymoron. The phrase makes perfect grammatical sense, yet the terms 'wild' on the one hand, and 'captive'/'captivity' on the other, do not yield conceptual coherence.

'Wild animals in captivity'

If posed the question 'what are zoo animals?', zoo professionals, zoo scientists and commentators on zoos in general are very likely to reply: 'Zoo animals are wild animals in captivity.' The term 'wild animals in captivity' is, therefore, not something dreamt up by this author as a 'straw notion' for the convenient purpose of exposing it as being conceptually flawed, but one which actually occurs in the titles and in the texts of many books on the subject of zoo animals – just to cite one classical example: *Wild Animals in Captivity: An Outline of the Biology of Zoological Gardens* by Heinrich Hediger, the famous director of Basle's Zoological Gardens, published in its English version in 1950.[1]

From the standpoint of the critique of this book, it is important to point out that there are three types of 'wild' animals the visitors could encounter in zoos, even though they might not necessarily be aware of the difference between them. They could be looking at a jaguar (*Panthera onca*) caught from the wild who may not have been a zoo resident for long; they may be looking at a jaguar, who has never known existence in the wild, having been born and bred in the zoo, although the offspring of one parent (or both) who has (have) been caught from the wild; they may be looking at a jaguar, the offspring of parents both of whom have themselves been born and bred in zoos, either as first generation zoo-bred jaguars or several generations down the line of zoo-bred jaguars.[2] It will be shown that the critique mounted against the term 'wild animals in captivity' applies to all these three categories of 'wild animal'. However, as the strongest case in defence of the use of the term 'wild animals in captivity' rests with the first type identified, it would be fair to focus on it and to assess the strengths of the arguments against its use even in such a *prima facie* favourable case.

Freshly caught 'wild animal in captivity'

At first sight, it seems very natural to consider the jaguar just caught from the wild and delivered to the zoo as a token of the species *Panthera onca*, characterised in Chapter 2, as a naturally occurring individual organism belonging to a naturally evolved species, by and large, within its historical habitat/environment. Surely, the very act of removing such an individual animal from the wild and relocating it within a zoo environment amounts to a peripheral matter which could not undermine the very meaning of the term 'wild'. Take an analogous case with regard to a human – the court has described the defendant in the dock as a dangerously violent man who regards women as objects to be abused, and has convicted him of a nasty rape, sentencing him in consequence to several years of imprisonment. The fact that he is now behind bars, under captivity, so to speak, does not render it (conceptually) inappropriate for society to continue to refer to him as 'that dangerously violent man who regards women as objects to be abused'. The phrase is perfectly intelligible; there is no incoherence in using it in the context of his incarceration.[3] If the analogy holds, then there should be no conceptual inappropriateness in talking about an animal being 'wild' and being 'under captivity' in the same breath.

However, the analogy fails to hold. The attribute 'wild' is not the same as the attribute 'weighing x kg', or the attribute 'large but sleek' in characterising a wild animal. The (male) jaguar weighing 48 kg (120 pounds) when caught in the wild, give or take, will continue to weigh around 48 kg when received by the zoo, although the stress of being caught and then transported to another location might make the animal lose a few kilograms in weight. But to simplify the argument: assume that the jaguar has not lost any weight, we can meaningfully talk about that 'jaguar in the wild weighing 48 kg' as well as that 'jaguar now in captivity weighing 48 kg' – in other words, being made captive or kept in captivity does not affect the meaning or the sense of 'weighing 48 kg' in any way, although as already mentioned, being captive could, as a matter of fact, affect the weight of the animal in question.[4] However, as we shall see, the logic of the words 'wild' and 'in captivity' are more complicated than 'weighing 48 kg' or 'having stripes all over the body'; we have already briefly drawn attention in the two preceding chapters to the deep themes in biology, especially of time and of the reciprocal causal relations occurring in the context of organisms-in-the-environment, which establish that an animal-in-the-wild cannot be properly and fully grasped except in the context of its history and its habitat within which it (as a species) has naturally evolved and in which it (as a member of the species) lives.[5]

'Wild' is the antonym of 'tame'[6]

We begin to unravel that complexity here by showing that the term 'captive' undermines and subverts the meaning of being 'wild' in a fundamental way. To demonstrate the subversion, let us turn our attention, in the first instance, to a meaning of 'wild' which is implicit in the exploration of animals-in-the-wild, though not explicitly dealt with in Chapter 2. An animal-in-the-wild, as a naturally occurring being, has no intimate contact with human beings, and would by instinct run away from such contact. Hediger has characterised the situation well and it may be worthwhile to quote him at some length; he talks about an

> irresistible impulse towards the continual avoidance of enemies, this flight tendency dominant in the animal's behaviour, and its manifestation in specific flight reaction. This is released by enemies, in the predator-prey relationship. Man often plays the part of the predatory animal, in fact there is hardly a species of animal that has not been hunted by him, often for centuries or even thousands of years.[7] Thus it may be said that man, with his world-wide distribution and his long-distance weapons represents the arch-enemy standing, so to speak, at the flash-point of escape reactions of animals.
>
> Yet not every approach of the enemy touches off the flight reaction, nor is every approach necessarily a threat. The situation only becomes dangerous when the enemy approaches to within a certain distance of the animal – the escape distance. Only when this specific flight distance, which differs for each species, is overstepped by an observed enemy does flight reaction follow; i.e. the animal in

a typical manner runs away from it, far enough to put at least its specific escape distance between itself and the enemy once again.[8]

(1968, pp. 40–1)

Given such a basic powerful impulse to run away from its perceived enemies, zoo visitors ought to ask themselves the searching question: why do these 'wild animals in captivity' appear not to do so? This is because they have been tamed, that is to say, they have been re-programmed by their human captors to suppress that basic capability, or even to eliminate it totally. Hediger says: 'Removal of escape tendency, i.e. taming and tameness, can only come from man. He is the only creature capable of freeing another from the magic circle of flight, from the irresistible impulse to avoid enemies continually' (ibid., p. 49).[9]

The taming process requires skills and expertise, but if properly undertaken the wild animal would gradually lose its flight tendency, permitting its human keeper/carer to approach it and under certain circumstances even to move freely in its presence. Without successful taming, zoo management is just impossible; unless anaesthetic darts are used every time to knock them out when they are examined for husbandry purposes, zoo carers/keepers must be able to approach them without provoking either a violent reaction (injuring or even killing the human carers) or a flight reaction. In any case, minimally, they have to be taught to get used to the presence of humans, as their captive environment is nothing but an environment which is run and controlled by humans.

Taming is essential not only for zoo animals, but obviously also for circus animals.[10] What is perhaps not so obvious, but which Hediger (ibid.) points out, is that it is a procedure essential to the evolution of domesticated animals or domesticants. Such animals cannot remotely be useful to us humans, unless we first tame and then ultimately breed out their tendency to escape – what good is a horse's pulling powers or the hen's capability to lay eggs, if the former cannot be harnessed and attached to a cart/carriage and the latter cannot be made to live in and lay their eggs in the hen house? In other words, taming and training are essential to the process of domestication itself.

This insight, in turn, then calls for an examination of the issue whether domestication, or at least some understanding of the notion, can be said to be appropriately applied to animals in zoos.[11] However, this matter is raised here only *en passant*; a later chapter will deal with it in greater detail.

Zoo-bred animals, whether first generation or not but whose ancestors are wild, still need to be tamed, as they, too, have to overcome flight reaction. However, with animals born in captivity or who have become habituated to captivity, the flight distance is somewhat less than in conditions under freedom in the wild. As we shall see, such zoo-bred animals have gone further down the line of 'domestication' than their counterparts just caught from the wild and freshly made captive in a zoo.

Conclusion

For the moment, let us sum up the main points which have been established in the light of exploring the conceptual links between the terms 'wild', 'tame', 'captive':

1. The phrase 'wild animals in captivity' is an oxymoron. This is because 'wild' as an attribute is not of the same kind as 'weighing 48 kg' in the context of captivity. For captivity to occur, it requires that the wild animal be tamed, that is, to lose its flight reaction and no longer to regard humans as the perceived predator which it would in the wild. 'Tame' is therefore the antonym of 'wild' (in one of the senses of 'wild'). While the phrase 'wild animals in captivity' amounts to an oxymoron, the phrase 'tame animals in captivity' amounts almost to a tautology, as to be a captive animal is to be a tame animal, and to be a tame animal is to be a captive animal in the context of zoos and their management. Empirically, it is also the case that without taming and training the animals, freshly caught from the wild as well as zoo-bred animals whose original ancestors are animals-in-the-wild, to respond to human ways of handling and relating to them, zoos cannot function and operate.

2. Taming amounts to a fundamental change in the behaviour of the animal taken straight from the wild or whose ancestors belonged to the wild. This change, programmed and orchestrated by its human captors and keepers, is so profound that it can be said to amount to the first but crucial stage of domestication. As a result, the gap which zoo professionals and scientists perceive to exist between zoo animals on the one hand, and ordinary domesticated animals such as horses or cows on the other, may not be as great as it is made out to be. They are at different ends of the same spectrum. However, as it is taken for granted that zoo animals are, unlike cows and horses, not domesticated animals, they are then mistakenly and wrongly called 'wild animals in captivity', implying in this context that 'wild' is synonymous with 'non-domesticated' and is the antonym of 'domesticated'. But as we have seen, at the level of surface grammar the phrase makes sense, but at the level of depth grammar it is incoherent, contradictory and hence unintelligible and senseless.[12]

4
Decontextualised and Recontextualised

The last chapter has argued that it makes no conceptual sense to talk of 'wild animals in captivity' by establishing a fundamental meaning of 'wild' in terms of its antonym 'tame'. This and the chapters following will go beyond the conceptual issue to unravel the nest of complexities behind the meaning of 'wild animal' and at the same time explore a crucial matter underlying those complexities, namely the ontological difference in status between animals-in-the-wild and their counterparts kept in zoos as captive animals. The difference may be explored in terms of the following basic aspects in the transformation of wild animals into what this book calls 'immurated' animals in zoos, that is to say, in transforming naturally occurring animals to become, to an extent (to be clarified in later chapters), biotic artefacts at the hands of *homo faber*:

1. Geographical dislocation and relocation.
2. Habitat dislocation and relocation.
3. Lifestyle dislocation and reaccommodation.
4. Suspension of natural evolution.

In other words, the animal caught from the wild, first becomes decontextualised and then recontextualised according to human design and intention. In particular, this chapter will consider the first of these two aspects. It will also begin to tease out the implications of the official definition of zoos as institutions which house collections of animal exhibits open to the public, in order to establish eventually that zoo animals are indeed the ontological foil to wild animals-in-the-wild.

Geographical dislocation and relocation

Chapter 1 has already mentioned that zoo professionals differ from lay persons in the connotation of the term 'exotic'. The latter regard an exotic animal as simply one which is unusual, strange, rare or from distant parts; the former understand the term to refer merely to animals whose natural home range is in a part of the world other than where the zoo which houses them is to be found, irrespective of whether they are charismatic or rare. In other words, in spite of the differences, the lay connotation does, to an extent, overlap with the technical understanding of the term – 'being from distant lands/parts' is perfectly compatible with 'whose natural home range is in a part of the world other than where the zoo is'. In the majority of cases, as a matter of fact, it would be true to say that exotic animals found in zoos are from distant lands/parts and, indeed, because of that fact are considered by the public to be unusual, rare, strange.[1] For instance, historically, most zoos have been found in Western and Central Europe, North America, Japan and Australasia. The *World Zoo Conservation Strategy* (1993) reckons that there are probably at least 10 000 zoos in the world, of which just 1200 could be said to be members of national or regional zoo associations – amongst this more exclusive group, North America counts 175, Europe 300, East Asia 545, Australasia 30, Latin America 125 and Africa 25.

In other words, zoos which have the most resources and are the best organised are zoos in the developed world which, by and large, are found in temperate climes, in the Northern as well as in the Southern Hemispheres, whereas the more exotic (in the lay sense of the term) and charismatic of the zoo animals, such as the lion, the rhinoceros, the chimpanzee, the elephant, the jaguar are also exotic in the technical sense. These have been transported to more temperate climes from their natural home ranges in tropical and savannah Africa (respectively the rhinoceros and the lion); in Western Africa, lying just above the equator from the coast to inland (the chimpanzee in tropical rainforests, woodlands, swamp forests and grasslands); in India (the elephant); and in the tropical forests of Central and South America (the jaguar).[2]

If it is correct that zoos deal with exotic animals, *ex hypothesi* it follows that its captive residents have been subject to geographical dislocation. The polar bear is no more a native of Australasia than the penguins are natives of Western Europe, or the elephant (whether Indian or African) of North America. As these animals have been wrenched from their natural home ranges, they have necessarily become decontextualised. Different geographical locations in the main mean different geographical features, different climates, different flora and fauna, different habitats, different environments *tout court*.[3]

Such exotic animals have to be recontextualised within a zoo setting. That is to say that zoo management has to provide them with a new setting. In the bad old days, their new setting was a cage; today's more enlightened zoo management provides them instead with an *ersatz* habitat and environment. The operative word is *ersatz*, as the most which can be offered is a naturalistic, that is, a simulated one, added to which is a programme of environmental enrichment.

While an examination of the notion of environmental enrichment will be deferred to the Appendix, that of a naturalistic environment needs to be looked at in brief details now.[4] However, before doing that, one must first focus on a fundamental point concerning the recontextualisation of such exotic animals.

Recontextualised as a collection of exhibits

In the wild, as we have seen, animals live 'for themselves' in following their own trajectories and unfolding their own *tele*. They do not live for us humans, no matter how fascinating we may find them, no matter how desirous we may be to understand more about them from the standpoint of pursuing scientific knowledge. But the moment they become exotic animals, they lose the status of existing only 'for themselves'; instead, they become incorporated into a human structure and a human project.

A zoo is a human social institution like a museum. Every museum has a collection policy regarding the type of artefacts it aims to acquire, in identifying gaps in its extant collection, and the like. While a costume gallery concentrates on acquiring garments belonging to a certain culture and a certain period in history, while a natural history museum collects stuffed birds, mammals and so on, a zoo exists to collect (a certain number of) exotic animals which are not stuffed, but alive and living. A collection in each of these three kinds of institutions just identified is necessarily a *collection of exhibits*.

According to *The World Zoo Conservation Strategy* (1993), what distinguishes zoos from other sorts of collecting institutions is precisely this: 'The fact that all zoos exhibit living specimens of wild animal species is what underscores the difference between zoos, most museums and other cultural or recreational institutions, and is what gives zoos their own unique character' (ch. 1.3).[5] It refers to 'different concepts in exhibition that play an important role in establishing the character of a zoo' (ch. 1.2).

Every zoo, therefore, necessarily possesses a number of exotic living animals which constitutes its collection of exhibits. By turning animals from the wild into exotic exhibits, zoos have robbed them of their onto-logical status as beings who live 'for themselves' and transformed them into beings who live for us humans, to serve our ends. As we shall see in greater detail later, they no longer live out their own trajectories and their own *tele*, but are constrained to exist in accordance with a plan, a purpose dictated by a power outside of themselves. This is to say that the thesis of intrinsic/immanent teleology has been displaced by the thesis of extrinsic/imposed teleology – as a collection of exhibits, they serve the ends which humans claim zoos are designed to advance, be these to educate (humans), to make them the object of scientific research and study, to save them from extinction in the wild, or provide entertainment by way of a day's outing for the family. Their value in the context of such a human project is, therefore, an instrumental one. In the wild, animals, while living 'for themselves' may, by happy coinci-dence, also have instrumental value for us humans; but in zoos, as part of a collection of exhibits, they assume (almost exclusive) instrumental value for their human captors/keepers/visitors.

A population of cheetahs or a population of antelopes in the wild is not a collection of exhibits – the population is not a collection, nor are its individual members exhibits. Similarly, a population of cheetahs and a population of antelopes coexisting in the wild are not a collection of exhibits; safari parties may set out in their specially constructed vehicles to roam their habitats with the explicit purpose of looking out for them, and if luck really holds, of even coming across a cheetah in hot pursuit of an antelope. The fact that such tourists can see them does not necessar-ily turn them into mere exhibits; in any case, these animals may even adapt their hunting habits to avoid the intrusion of such human voyeurs by no longer hunting during the day but at night, or at a time when these unwelcome lookers-on are not about. But if such animals are captured and relocated to zoos, then their *raison d'être* is precisely to be visible, to be seen by humans, at times predetermined by zoos. If certain animals do not, alas, keep zoo opening hours, they are made to do so – nocturnal ani-mals, in so-called nocturama, are made to come out 'at night', to be active, to hunt for the benefit of zoo visitors, and made to rest, to sleep during 'the day' with the help of ingenious artificial lighting.

A population of cheetahs in the wild, whose members hunt, rest, groom, mate, reproduce, look after and teach their young survival skills so that in turn they can grow up to hunt, to reproduce, to transmit knowledge to their offspring, is from the evolutionary point of view an important bearer of information, genetic, ecologic and cultural. But the living out of their existence and the transmission of such knowledge have nothing to do with human intentionality of any kind, and most certainly not with the particular human project of rendering them conveniently and suitably visible to the human gaze.

Animals-in-the-wild, as individuals or as species, are simply not exhibits. However, once made captive in a zoo, they become exhibits, much as Leonardo da Vinci's *Mona Lisa* hanging in the Louvre is an exhibit, much as the skeleton of a dinosaur hanging in a natural science museum is an exhibit. The difference between a stuffed polar bear and a polar bear in a zoo is that one is a dead organism and the other is a living organism – they both have labels, so to speak, hanging round their necks and are displayed to advantage for the benefit of those who visit natural science museums or zoos. Just as one would design special cabinets, cases, installations to show off important exhibits in an art museum, one would do the same for certain animals in zoos. The bare cage is not necessarily the only way by which one may exhibit the animals. Elaborate structures have been built to show them off 'at their best', according to zoo-speak. These clearly embody a desire not merely to put such animals under human control and management but also to incorporate them into the culture of their owners and controllers. In 1856, at a time when Europe was particularly gripped by the Pharaohs, their rediscovered tombs under the pyramids, not to mention the deciphering of their hieroglyphs, Antwerp zoo, for instance, built an Egyptian temple whose walls were decorated with frescoes and bas-reliefs depicting scenes from ancient Egyptian arts to display its elephants, giraffes and camels. Also popular were Moorish structures and Asian pagodas. Budapest zoo even went in for the Byzantine style of architecture when it built its elephant house, complete with Byzantine church interiors.[6]

One zoo expert even goes so far as to say that modern zoos present spectacles based on a combination of the techniques of museum exhibition and theatrical performance.[7]

> exhibits are designed to imitate the natural habitat of the species, which will enable the animals to express natural behaviours with the zoo environment. In some respects modern zoo design may be compared with a theatre where the animals are the actors, the exhibit design is the scenery, the theme or the story being interpreted is the play, the visitor area is the auditorium and the visitors are the audience.
>
> (Andersen, 2003, p. 76)

Of course, this insight is absolutely valid; however, it is proffered by the author with no apparent awareness that such a context is incompatible with the implicit claim that the animals under display continue to enjoy the status of 'being wild' animals expressing 'natural behaviours'.[8]

Habitat dislocation and relocation: the naturalistic environment

In the wild, as we have seen, animals live 'for themselves' at their own pace, in their own ways, under circumstances of their own choosing in habitats within which they have historically evolved and lived; in zoos, willy-nilly, they have to live for humans, no matter how enlightened the philosophy of zoo management regarding the conditions under which they are exhibited, as exhibited they must be. In other words, they have to live under conditions not of their own choosing but under those designed and chosen for them by their human keepers/carers.

Today, enlightened zoos endorse a philosophy of management which wants nothing to do with the oppressive image of the bare and barren cage. Instead, they prefer to allow the lion to move about within a naturalistic environment. What, then, is such a space? The very word 'naturalistic' says it all; it is a space which has been designed and engineered to 'look natural' within which the animal can be displayed to better advantage as an exhibit. This is its fundamental *raison d'être*; however, to say this is not to deny that other justifications have been put forward. For instance, enlightened zoo management says that it is intended to improve the quality of life of the animals, which undoubtedly it does to a limited extent, or that such a setting would provide better educational opportunities for the visitors. Whether these claims are mutually compatible remain to be seen.[9] However, as we shall see, the latter two are at best secondary justifications and cannot subvert and displace the fundamental one of maximising the conditions under which visitors could appreciate the animals which are and must be seen as exhibits.

It is instructive to quote at some length what an expert on the subject of animal exhibit design has to say which corroborates the analysis above:

> The foremost goal of most zoological institutions is to use the entertainment value of live animals and re-created foreign worlds to draw people into an educational situation, in which they will learn about

and gain respect for the animals they are observing, as well as nature in general. Encouraging visitors to lengthen their stay in the park is important to increase these educational opportunities, as well as increase the institution's income (from concessions sales, for example) so that it can continue to thrive and carry out its work.

Visitors spend the most time at an animal exhibit when the animals are close-up and active . . . Most animals however, prefer to keep their distance from their human observers, if space allows. When space doesn't allow, i.e., they are in a small enclosure with no visual cover, these animals suffer psychologically, resulting in their being inactive, or in their performing unnatural and undesirable behaviours. Unnatural behaviours detract from the educational value of the exhibit, as they do not accurately represent wild animals. The sight of an unhealthy or unhappy animal will not instil in visitors a sense of awe for the natural world and may contribute to a poor reputation of zoos.[10]

However, give animals vast, heavily planted spaces in zoos, in which they can lead a natural lifestyle in relative privacy, and visitors will not be able to benefit from observing them, as the animals will often be out of view. This is why designing animal exhibits that are truly based on wild situations, and considering only the absolute needs of the animals, is often secondary in zoo exhibit design.

(Worstell, 2003, Introduction)

Furthermore, the provision of naturalistic environments must also be considered from another angle, namely, that it entails a severe monetary cost, which has to be taken into account, thereby restricting the construction of such designs mainly to exhibition areas while excluding off-exhibit (off-stage, behind-the-curtains) spaces and laboratory enclosures.[11] It is refreshing occasionally to find such a frank admission amongst certain zoo professionals such as the one cited below:

Animals on exhibit in zoos are presumed to benefit from the environmental complexity provided by naturalistic habitat enclosures . . . If the naturalistic exhibit also provides educational benefits for zoo visitors, then – so long as the enclosure does not compromise the residents' health – the time and effort may be justified. For laboratory enclosures or off-exhibit areas of zoos where every effort to enrich has a monetary cost, this may not be true.

(Crockett, 1998, p. 130)

The above simply confirms once again that the underlying fundamental assumption in today's enlightened zoo management philosophy is that animals in zoos are first and foremost exhibits.

Every design has to satisfy three basic requirements – comfort of the animal, the distance between the animal and the visitor, and habitat simulation – which are to be reconciled in the following way:

> Small enclosures, even when appropriate for the size and number of animals on display, will appear as a cage rather than a natural environment, unless they possess the illusion of being an undeterminable space. . . .
>
> In the design of animal exhibits, the optimal size for the animal enclosure achieves a perfect balance between animal comfort, animal proximity to visitors, and the portrayal of an essence of habitat. However, the relationship of the animal to the visitor, as well as enclosure features, influence the perceived space. The perception of the enclosure size by visitors and animals will be influenced by visual illusions and psychology.
>
> (Worstell, 2003, ch. 2).[12]

According to this way of thinking, the zoo designer tricks the human visitors into believing that the space is infinitely larger than it is in reality. Worstell (ibid.) shows a photo of the gorilla space at Burgers Zoo at Arnhem, Netherlands, which she says 'appears to be endless'. Endless to the human visitor, yes, but would it also be so to the animal which inhabits that space? Zoo professionals may be wrong in the assumption that the illusion works for both humans and animals. The former stand outside the enclosure and gawp at the animal within the enclosure. Visual illusion works better when the perceiver is at a distance and when the perceiver never gets any closer to what is perceived. Travellers in a desert report the mirage phenomenon, that they see water in the distance. However, we and they know that this is an illusion and not reality, for the simple reason (apart from the scientific one) that when the travellers get to the very spot where the mirage occurs, they will find that there is no water. One can only happily take illusion for reality if there is no other reality check available. The zoo visitors are exactly in such a position; they can live happily with the designed illusion that the space occupied by the exhibit is 'endless'.[13] The gorillas, however, are like the travellers in the desert and they would surely know, after the initial deception, that the space they occupy is not 'endless'. They have reality checks not available to the human visitor standing outside the enclosure. Their initial visual perception would be very quickly corrected once they start to stroll around the enclosure. So, who is the deceived party? It is surely not the inmate of the enclosure, but the human visitor whose perception has been deliberately manipulated by zoo designers.[14] After all, it is precisely the job of the designers to produce under conditions of simulation just such an illusion, in order to maximise the opportunities for exhibiting the captive animals – zoo visitors no longer want to see or derive maximum satisfaction from seeing animals in cages but prefer to see them in a naturalistic environment. Habitat simulation which consists of 're-creating the essence of a natural habitat' is considered to be essential to the visitor experience.[15]

To simulate, let us say, an African tropical rainforest, the natural habitat of gorillas, the presence of trees in the exhibit enclosure is essential. However, maintaining live trees may be costly; furthermore, the animals love to peel the bark and may destroy the trees. So the ideal solution – that is, from the standpoint of zoo management – is to construct artificial, virtually indestructible trees which from a distance look as convincing as the real stuff with which they may be mixed. The public who is ignorant in general of such ploys is happy with the visual effect and that is what counts. After all, '[a] visit to the zoo is primarily a visual experience' (Worstell, 2003, ch. 5). The gorillas would not be duped but that is another matter.

In temperate climes, tropical vegetation may present problems. But one can, like Munich Zoo, for instance, select evergreen grass, which is native to Spain, as the cover for the natural ground of its primate exhibits. African rainforest plants have leaves which look like those of laurel (*Prunus larocerasus*). So laurel would do, as it possesses desirable characteristics such as large, shiny, year-round foliage; however, it is not recommended for use in primate exhibits as it is toxic, although Chester Zoo has successfully used it in its chimpanzee exhibit where the chimpanzees appear to avoid eating it. Temperate needle evergreens do not remotely resemble any plant found in African tropical rainforests, but they are recommended all the same for gorilla exhibits in order to create a green landscape, especially useful during the winter months, as these animals, on the whole, do not destroy them; however, plant them right at the back of the exhibit to make them less obvious to zoo visitors, just in case they recognise that they are natives of temperate, not tropical, forests. To sum up briefly this line of thinking, according to Worstell (2003, ch. 6):

[i]t is not important that gorilla exhibits in temperate zones cannot be faithfully planted with rainforest plants. What is important is that the essence of the gorilla's natural habitat is conveyed. This can be

done with hardy plants, as habitat replication depends more on the character, growth habit, arrangement, spacing, massing and diversity of plants, rather than the actual species.

Habitat simulation is a perfectly coherent ideal within the context of exotic animals as a collection of exhibits for exposure to the human gaze. The average zoo visitor is not a botanist with profound knowledge of different habitats and ecosystems in the world. However, it may be true that the public knows this much, namely that while many trees (deciduous) in temperate climes shed their leaves in the autumn, evergreens do not; that in tropical climes, plants on the whole do not shed their leaves in the decisive manner which deciduous trees do in temperate latitudes; that trees in a tropical forest tend to grow straight and tall, throwing up a broken canopy under which shrubs grow. It is this minimal knowledge about plants and trees in different parts of the world which the zoo designer is exploiting, as evidenced in the discussion above. The simulation reinforces this general, scanty botanical knowledge while at the same time providing a pleasing illusion to the public that they are in the presence of the natural habitat of the animal which constitutes the exhibit.

Judged by such a measure, habitat simulation, properly constructed, may be a commendable success. However, more than this is claimed for it by its advocates. They wish to argue, as we have seen, that it is the re-creation of the 'essence of a natural habitat'. We have already referred above to enough details about its conception, design and construction to make the obvious point that simulation is the essence of a naturalistic environment.[16] A simulation of something could not by any stretch of the imagination be construed as re-creating the essence of something, unless it is assumed that the 'essence' of something is nothing more than what it appears to be at a distance. This is analogous to an art historian claiming that by digitalising Michelangelo's *David* or any other aesthetic artefacts, making the images available on its web page, allowing the viewer to rotate the images so that the back of the object can also be looked at, that such an attempt amounts to 're-creating the essence of the art object'. No reputable art historian, to one's knowledge, has made such a claim, which would only make sense if visual perception of an electronic kind alone counts as knowing the object. Museums do not boast that virtual reality captures the essence of the art object; virtual reality is no substitute for the real presence.

It is true that museums, on the whole, do not allow visitors to touch their exhibits, although it may designate a very limited number for touching and holding under supervision, and it may also allow students of the subject to have greater access to them for the purpose of teaching appreciation of the artefacts along with students acquiring more profound knowledge of the objects of their study. 'Do not touch' is the general rule for the public for the simple, valid reason that art objects are rare and handling by all and sundry is a recipe for disaster and destruction, whether short or long term. It is *faute de mieux* that museum visitors are confined to only looking at the exhibits; museums do not claim that the gaze – whether by actually visiting a museum or by calling up its contents on its sophisticated web sites which permit multi-media interaction – is all, though it remains true that the gaze is better than no gaze, should one wish to appreciate the beauty and complexities of artistic artefacts. It is true that the gaze in the presence of the artefact in a museum works best for paintings. However, in the case of artefacts such as statues, pots, armour, swords, and so on, these can only be fully appreciated not simply via the visual sense, but also the kinaesthetic, and indeed sometimes even the auditory sense, as in the case of (what the West calls) true porcelain which, when 'pinged', gives a distinct sound, a sound which forms part of its 'essence'.

Visual perception is, therefore, only one mode of exploring the world and acquiring knowledge of it, whether the agent is a human animal or a non-human animal (like the exotic ones which we see in zoos). It is true that we, humans, rely more on visual perception than other animals. One ascertains what is out there in the world by looking, and not so much by touching, smelling or tasting. Humans, in comparison to other animals, have less acute auditory and olfactory senses; furthermore, in leaving the hunting and gathering mode of existence behind them, to become, so to speak, civilised, these senses have become even more diminished in the course of human evolution.

Furthermore, it is also true that in modern epistemology, the visual is privileged over all the other perceptual modes. By the time the age of modernity arrived (roughly dated to the seventeenth century, as far as Western Europe is concerned), verification *via* observation, using the eyes either directly or with the help of instruments (such as microscopes, telescopes) became the dominant mode of appropriating the world and of gaining knowledge about it. Modern science celebrates visual perceptual knowledge, as it appears to satisfy the scientific methodology of objectivity; objectivity takes the form of measurement and quantification of data. Sight in this broad sense is considered to be more reliable as a source of knowledge because one can measure objectively what one sees – one can determine the length, width, height and weight of the table we see in front of us with the help of instruments, such as a tape measure, a weighing machine. Touch is unreliable except in a situation where an instrument, such as a thermometer is available to measure the temperature of the liquid, and so on. Taste is a mere 'secondary quality', said not to reside in the object itself (unlike length, breadth, weight, shape, which are 'primary qualities', are real and reside in the object), and a singularly subjective one. As such, it falls outside the domain of science.[17]

The claim made on behalf of habitat simulation appears to carry the above reductionist epistemology to an absurd extreme. The 'essence of a natural habitat' cannot be captured, first, simply through visual perception, and second, through the visual perception of a simulated habitat. If it could be thus captured, it would follow that the exotic captives are living in their natural habitats, as opposed to being merely perceived as such by their human visitors in an *essentially human cultural/institutional* setting and context, exclusively contrived by their human captors/designers to frame them *essentially* as *exhibits*.[18]

Conclusion

This chapter has argued that ontological dislocation follows from geographical dislocation and relocation, as well as from habitat dislocation and relocation via habitat simulation, when exotic animals are displayed as collections of exhibits in zoos. They are no longer naturally occurring beings which live 'for themselves' in their natural habitats, but beings which primarily have instrumental value for their captors/keepers/carers/visitors as exhibits.

5
Lifestyle Dislocation and Relocation

The last chapter has critically examined two aspects – geographical and habitat dislocations – in the ontological transformation of naturally occurring living beings (wild animals) to become exotic captive beings, relocated to a different setting and framed within a context deliberately designed by humans for the purpose of displaying them as exhibits. This chapter continues to delineate that transformation by looking at the dramatic changes to the lifestyle of wild animals when they are made part of a collection of exhibits in zoos. Their ontological implications will be teased out under two further aspects:

1. Spatial miniaturisation in habitat simulation.
2. Hotelification.[1]

Spatial miniaturisation in habitat simulation

It is obvious that today's enlightened zoos do not, on the whole, confine and exhibit their animals in a bare, concrete cage; instead, as we have seen, they exhibit them in a naturalistic environment or simulated habitat as part of their philosophy of environment enrichment. So is it fair, then, to talk of spatial confinement when their enclosure is no longer the mere size of a cage? This section will argue that it is; the argument begins by bringing certain salient facts before the reader.

Exotic captive animals in zoos naturally vary in size, although, on the whole, it is fair to say that, as far as mammals are concerned, they are not usually as small as the size of say a rodent or a bat.[2] So let us first focus on medium and large mammals, – carnivores and herbivores – as they exist in the wild, and highlight some of their distinctive characteristics in their original natural habitats.

The cheetah (*Acinonyx jubatus*) is different from the other cats in that its sleek body has evolved for speed, although the claim that it could do 68 mph (110 kmh) remains unsubstantiated.[3] The male weighs between 50 and 72 kg and the female between 35 and 63 kg. They are found in open habitats such as grasslands and semi-desert, but never in dense forests. The Indian cheetah became extinct in the twentieth century; today, remaining significant populations are found in Central and East Africa. The home range of female cheetahs is extensive as they may roam over 800 km^2; in the Serengeti, they follow the annual migration of the Thompson's gazelles. Males travel less extensively; they form coalitions numbering two to four, marking territories about 40 km^2. However, in the Kruger Park in South Africa, there is no Serengeti-like migration; females hold territories similar in size to the males which do not form coalitions.

The gorilla (*Gorilla gorilla gorilla*) is found in Africa, from south-eastern Nigeria to western Zaire and eastern Zaire into adjacent countries. The habitat of highland gorillas is up to 3000 m (10 000 ft) above sea level; western lowland gorillas inhabit secondary forests with widely spaced, slender-trunk trees with a broken canopy, permitting sunlight to penetrate; such landscape features make it easy for the gorillas to move about on the forest floor, as they are primarily ground-dwelling animals. They move about in troupes with a dominant male, travelling extensively. The troupe's home range is between 10 and 40 km^2 (4 and 15.35 mi^2); the group travels up to 5 km per day. Its nomadic lifestyle means that it is subject to changes in its surroundings on a daily as well as on a seasonal basis.[4] The adult male weighs 135–230 kg; the females are smaller.[5]

There are three main species of elephants in the wild: one Asian (*Elephas meximus*), two African – African savannah (*Loxodonta Africana*) and African forest (*Loxodonta cyclotis*). Most zoo African elephants belong to the African savannah species. The fully grown African bull can weigh between 14 000 and 16 000 lbs (6300 and 7300 kg) and grow up to 13 ft (4 m) at the shoulder; the Asian elephant is smaller, the average weighing 5000 lbs (2300 kg) and is 9–10 ft (3 m) tall. The distances travelled by the herd depend on the availability of food – when food resources are scarce, African elephants cover vast distances of several hundred kilometres. However, the median herd home range is (was) 113.0 km^2 for Asian and 1975.7 km^2 for African elephants. The Asian herds travel daily on average 3.2 km while the African ones cover 12.0 km. (Home ranges of 10–800 km^2 have been recorded for Asian elephants and 14–5527 km^2 for the African elephants.) The male, when in must (which can be at any time during the year), roams extensively looking for receptive females.[6]

Polar bears (*Ursus maritimus*) are the largest land predators. The species is the only one out of eight others whose diet is almost entirely carnivorous. The polar bear is about four foot tall and weighs 400–1700 lbs, with the male being much larger than the female. It lives on broken ice packs off the northern continental edges near the North pole (but not generally above 82° latitude), a region called the circumpolar Artic, which covers Canada, Alaska (USA), Russia, Norway and Greenland (Denmark). In the winter, the bear dens and the mother bear produces her cubs. However, the polar bear is not in true or deep hibernation, although its heart rate slows down, its temperature falls a little, and it stops urinating and defecating. Its sense of smell is acute, enabling it to detect prey – it can smell a seal more than 32 km (20 miles) away. It is extremely well insulated, not merely with its fur but also with a layer of blubber which can be as thick as 4.5 inches; as a result, it gives off no detectable heat and, therefore, cannot be picked up by infra-red photography. It is an excellent swimmer, paddling at 6.5 mph; it can swim for several hours at a time, and some polar bears have been tracked swimming non-stop for 100 km (62 miles). Polar bears move throughout the year within their individual home ranges, which can vary in size, depending on access to food, mates and dens. A small home range may cover 50 000 to 60 000 square km (19 305/23 166 square miles); a large one may be over 350 000 square km (135 135 square miles). Polar bears can travel up to 30 km (19 miles) a day for several days – one polar bear managed 80 km (50 miles) in 24 hours and another 1119 km (695 miles) in one year. The female with cubs hunts about 19 per cent of her time in the spring and double that in the summer, while the male hunts about 25 per cent of his time in the spring and about 40 per cent during the summer. Polar bears are on the whole solitary beasts, with two main social groupings – adult females with cubs and breeding pairs.[7]

A recent study based on a comparison of the home ranges of wild mammals in their natural habitats and the enclosure sizes of the simulated habitats of their captive counterparts in UK zoological collections has come to the following significant conclusion:[8]

This (sic) results show that the bigger the animal the bigger the difference between its enclosure size and its natural home range. . . . megafauna . . . are kept in UK zoological collections in enclosures an average 1000 times smaller than their minimum home range. . . . The average difference for all the mammals study (sic) suggest (sic) that mammals in the UK zoological collections are kept in enclosures about 100 times smaller than their minimum home range. Considering these values, and applying it to human beings, a person that has lived in a village of 1 km² most of his/her life would be in the same spatial situation than (sic) a captive zoo animal if this person was (sic) confined for life to live in a telephone box. . . . 10 per cent of the taxa show enclosure sizes similar than (sic) the minimum home ranges, more than a third of the taxa . . . 10 times smaller, more than a fifth 100 times smaller and 2 per cent 10 000 times smaller. There are no cases, though, where average enclosure size was bigger than minimum home range. . . . It is to notice that these results only take in consideration area, not volume. Many species live in habits where the third spatial dimension is very important (tree dwelling, for example), so although they might be living in an enclosure of roughly the same size as their Minimum Home Range (like 10 per cent of the taxa of this study), they may lack the third dimension, and therefore they may still be restricted. . . . Our results show that the actual elephant enclosure sizes in the UK, as opposed to the (sic) recommended by zoo organizations, are in fact an average of 1000 times smaller. . . .

Mammals kept in UK zoological collections during the period 2000–2001 are confined in enclosures that, as an average, have an area 100 times smaller than their minimum home range.

Mammals with a body mass bigger than 100 kg (Megafauna) kept in UK zoological collections during the period 2000–2001 are confined to enclosures that have an average area 1000 times smaller than their minimum home range.

(Casamitjana, 2003, p. 13)

Medium-size mammals, such as the cheetah and the female gorilla are unlikely to fall into the 10 per cent of the taxa whose enclosure sizes are the same as those of the minimum home ranges. Of the elephant, Clubb and Mason have this to say:

A survey of Asian elephant enclosures in 20 European and North American zoos show(s) a range of 100 to 48,562 m², or 17 to 4937 m² per elephant . . . wild elephants roam over considerable distances. Even considering minimum wild home range sizes, the outdoor enclosures recommended by both the EAZA and AZA are in the region of 60 to 100 times smaller.

(Clubb and Mason, 2002, p. 41)

These figures concern only the size of the outdoor enclosures. One must, however, bear in mind that the average elephant in northern zoos spends up to 16 hours a day during certain periods of the year in indoor enclosures which are naturally much smaller than the outdoor ones whose major, if not sole, purpose is that of exhibition. In other words, these elephants have to spend a large proportion of their time within a space of just 36 to 45 m². We have also seen that the Asian elephant travels on average 3.2 km and the African elephant 12 km, and can cover as much as 17.8 km a day. No data are available for the distance covered by the zoo Asian elephant; the best figure available for African elephants in zoos is an average of 3 km a day, which is very much below what the wild elephant in Africa achieves.[9]

Mason and Clubb (2003) have also done a more recent study (funded by the Universities Federation for Animal Welfare and six British zoos) based on an analysis of about 1200 articles published in learned journals during the last 40 years on observations of animals-in-the-wild and 500 zoos world-wide; in particular, they have this to say about polar bears, that the typical zoo enclosure is one millionth of the size of their natural home ranges.[10] It may be pertinent to point out, too, that Central Park Zoo, relative to some other establishments, can be said to be generous in providing room for its polar bears, as they are allowed a 5000 square-foot exhibit space; however, it is, nevertheless, pertinent to bear in mind that the average natural range of polar bears in the wild is 31 000 square miles. Polar bears on the move can travel, as we have seen, 30 km (19 miles) a day. To achieve that distance, the captive polar bear in Central Park Zoo would have to do many a turn in that 5000 square foot enclosure in order to cover 19 miles, just as a prisoner in a small cell would have to pace up and down its length and breadth endlessly in order to get a decent dose of exercise to keep fit. But as we shall see in a minute, unlike the prisoner who may have a strong motive to keep himself perambulating within the confines of his cell walls, the captive polar bear is deprived of any major motivation to roam or move about in an analogous manner. Central Park Zoo as part of its philosophy of environmental enrichment also provides a pool for the bears to submerge themselves in. At the back of the exhibit enclosure is an ice machine which produces piles of ice to imitate Arctic conditions. To an animal capable of swimming and travelling as much as 30 km (19 miles) a day, and to cover as much as 259 000 km² (100 000 mi²) in a life-time, a pool ensconced within an exhibit space of 5000 square foot would be miniaturisation carried to extreme. Ironically, should the animal wish to travel the same distance per year as it would do/have done in the wild, it would have to go round and round the pool non-stop for most days in the year, a stereotypic phenomenon considered to be distinctly pathological.

Hotelification

The term 'hotelification', coined by this author in this context, means no more than what the word 'hotel' normally denotes and connotes. A hotel exists to provide two basic services to a traveller, namely accommodation and food. A zoo also provides these two kinds of services to the exotic animals which have been transported a long distance away from their natural habitats. The only difference between the human and the animal traveller is that while the former usually stays for only a short or limited period of time in a hotel, the latter and their descendants are made to stay in 'hotels' usually for the rest or the whole of their lives, with rare exceptions where a small number of them are returned to the wild.[11] 'Hotelification' denotes the procedures of incorporating exotic animals in captivity within the facilities of lodging and full board offered by zoos.

Animals-in-the-wild spend a lot of their waking time looking for food and a safe niche to bed down as well as, in the case of females, to give birth to their young. The polar bear is an excellent example of this latter need. Mating is some time between late March and mid-July. The fertilised ovum divides a few times and then implants itself in the uterus for up to six months – this is called delayed implantation. Around September, the embryo attaches itself to the uterine wall and starts developing again. The mother polar bear dens in October or November and the cubs are born in December or January while she is still hibernating. The cubs are born blind, hairless, toothless and very small, weighing only 600–700 grams (21–25 ounces). However, within several weeks, they would have grown to 10–15 kg (22–33 lbs) feeding on the extremely rich milk of the mother. In the case of the cheetah, the cubs are also born blind and toothless; it takes five weeks before their eyes open. The need to find a safe niche for them is obviously imperative.

Another imperative need for animals-in-the-wild is the search for food. As we know, cubs have to be patiently taught where to find, how to identify as well as how to obtain it. It takes a cheetah mother up to two years at least to teach her cubs to master the art of hunting/stalking before they can be left to hunt on their own. The cubs have to be taught to perfect a carefully choreographed stalking-cum-killing sequence known as chase-trip-bite. A similar period of apprenticeship is involved in the case of the polar bear cubs whose mother has to teach them what to eat, where and how to catch the prey, and in the case of females where to den, as well as to negotiate the many dangerous hazards facing them on the ice.

Medium and large mammals in the wild tend to roam varying distances each day primarily because of the imperative to seek food.[12] As they are perambulatory, they often, if not invariably, have to seek out new shelters each day. As a result, their waking moments tend to be taken up with three basic activities, namely travelling, feeding and nesting, leaving no time to play.[13] For instance, the gorilla spends about 30 per cent of its daily existence travelling, 40 per cent resting and sleeping (gorillas in the wild build their nests on the ground, building their nests afresh daily, choosing carefully a selection of plant material), and 30 per cent foraging. Gorillas are primarily vegetarian, although western lowland gorillas also eat large quantities of termites and ants. Their diet in the main consists of more than 70 different plant species which include bamboo, wild celery, vines, berries, roots and bark. A male can eat up to 75 pounds of bamboo a day and a female up to 40 pounds.[14] Elephants in the wild dine off a wide range of plants (the Asian elephant ingests up to 75 species), mostly grasses but also bark, twigs, leaves, roots, flowers, herbs and fruit, depending on the season; given the low quality of the vegetation and their sheer size, they naturally have to spend 60–80 per cent of waking hours feeding to get what is needed for their nutrition, consuming between 150–300 kg of (wet weight) forage in the case of adult elephants. The Asian elephant is very adept in using its feet to dig up roots and to scrape vegetation lying close to the ground. Elephants also use their trunks to great effect as a means of removing dirt or thorns from their food.[15] And as we have seen, the distances travelled are entirely dependent on the availability of food; the African elephant roams more extensively than its Asian counterpart because of this factor. The cheetah preys primarily on other mammals weighing less than 90 pounds, particularly impalas and Thomson's gazelles. It also goes for springbok, steenbok, duikers, hares, the young of warthogs and wildebeest, kudu, hartebeests, oryx, roans, sables and even birds.[16] The polar bear lives mainly on ringed seals and bearded seals; depending on availability, it also goes for harp and hooded seals. It would also dine on fish, birds, bird eggs, small mammals, shellfish, crabs, starfish, dead animals (like dead whales, walruses, narwals), mushrooms, grasses, berries and algae. In late May, when the seals emerge, it gorges on them; an adult needs to gain 100–200 pounds in weight before the ice breaks up and the seals leave the area. The polar bear uses a variety of methods to catch its prey, such as still hunting, stalking on land, aquatic stalking and stalking birth lairs (of seals).[17]

It is obvious that zoo diets for its captive animals are totally different from those which the animals-in-the-wild enjoy. Zoo gorillas (at least in

US zoos) are fed fruit, vegetables, monkey chow, nuts and seeds.[18] Captive elephants are fed dried forage (mainly timothy hay), some commercial concentrate feed in the form of pellets, a small amount of fruit and vegetables, vitamin and mineral supplements, and sometimes branches and leaves (browse), the last depending on availability. They have virtually no access to pasture grass. Given the higher nutritional value of the zoo diet, they would not have to eat as much in terms of volume of food compared to their counterparts in the wild.[19] In US zoos, the national zoo diet for cheetahs consists of ground horsemeat and sometimes beef (made up especially for zoo carnivores), rabbits and chicks. Polar bears are fed 'omnivore chow, four to five pounds of fish daily, and apples and carrots'.[20]

Zoo diets for captive mammals (medium and large) differ fundamentally from those which their counterparts in the wild pursue in the following ways: although by and large equivalent in nutritional value, the items of food in zoo diets bear little or no resemblance in phenomenological terms to those available in the wild.[21] This is to say that in terms of appearance, taste, texture, smell, the two sorts of foods are essentially different. Furthermore, the ways in which zoo foods are served up are necessarily not foods which the animals themselves have hunted or foraged – zoo animals enjoy 'room service' and 'table d'hôte'.

Conclusion

The spatial miniaturisation of simulated habitats and the services of hotelification may be necessary and indispensable components of zoo management; however, it remains reasonable to argue that these two managerial techniques have the effect of inducing an 'existential crisis' in captive mammals (at least of the size that this chapter focuses on). One must straightaway enter a philosophical caveat here – talk of an 'existential crisis' should not and need not be understood in the same way as such talk would in the human context. The term here is only used analogously. Humans, as we know, have a peculiar type of consciousness which is characterised by a capability for abstract thought, for symbols, for language in general. The linguistic dimension renders humans capable of deep reflection about life in general and metaphysical speculations about the meaning of life in particular. For humans, an existential crisis is usually a crisis of identity. Therefore to say that captive mammals in zoos suffer from a crisis of identity is not to imply that they do so in the same way humans suffer from it. Human captives sent to the Gulags can ponder and reflect, as well as articulate their thoughts and feelings under their changed predicament; captive mammals cannot do that, as they lack the unique type of consciousness that humans possess. However, this is not to say that they – those recently captured from the wild as opposed to those who are entirely zoo-born and zoo-bred – are incapable in some ways (short of linguistic articulation) of noting their changed predicament, of their loss of freedom in doing what they would normally do if they were not removed from the wild.[22] To travel, to roam (whether nearer or further afield in the wild) in the search for shelter, for food (and later as we shall see, also in some cases for mates) in the precise ways each species has evolved over geological times constitute the existential predicament of the individual who is a token of their type (species), and deeply informs, therefore, the identity of these individuals as tokens of their respective types. That identity is undermined and indeed, totally dissolved under captivity, where the natural modes of choice and appropriation of foods are suspended and replaced by 'room service' foods in a 'hotel' with full board and lodging under human management – such a drastic change in circumstances may no less be characterised as an 'existential crisis' or a 'crisis in identity', which constitutes a major part in the ontological transformation of their status from that of naturally occurring beings to that of exhibits in zoo collections. In other words, they are subjected to humanisation, being drawn into the orbit of human culture, cared for, controlled and manipulated by humans for human ends, be these the high-minded goals of education and research, or the perceived less high-minded one of recreation/entertainment. It is none other than to be deprived of their status as beings which live 'for themselves', to manifest their own *tele* (given to them by the fact they each are tokens of the respective species to which they belong), to become beings which live essentially for the sake of humans. It is to lose their independent value as naturally occurring beings to assume, by and large, instrumental value only for humans.[23]

6
Suspension of Natural Evolution

The last two chapters have looked at the ontological transformation of animals-in-the-wild as naturally occurring beings to become captive exotic exhibits in zoo collections under four aspects, namely, geographical and habitat dislocations, spatial miniaturisation of their simulated habitats and hotelification. This chapter will continue to explore in the same spirit, first yet another aspect, namely, medication, before it goes on to show that all these arguments, marshalled thus far, justify the claim that zoo management of animals amounts to the suspension of natural evolution, and hence that the products of such management would, therefore, no longer qualify to be naturally occurring beings, given the change in their ontological status.

Medication

It is not unknown that some animals-in-the-wild take steps to get rid of unpleasant/painful symptoms which they feel in themselves; this is even to say that sometimes they take steps to treat and cure themselves.[1] Cats are known to go for cat mint as a carminative and:

> Costa Rican capuchin monkeys rub a plant resin into their fur that repels insects; wild pigs in Asia prophylactically consume plants with anthelminthic properties; and certain birds maintain intestinal regularity by ingesting tomato-like fruits that contain a natural laxative.[2]

However, this is not to say that animals have a concept of medication which is equivalent to the one we have in (human) culture. They have no shamans, certainly no professional doctors and paramedics, no theories of illness/diseases, no hospitals, no machines whether high or low tech, no refined drugs produced by a pharmaceutical industry to sort out their ailments. Animals-in-the-wild succumb to illnesses and diseases from which they eventually die. No animal can survive should they hurt themselves in any serious manner – they have to be sound in their limbs, not to mention in all their organs, to be mobile in order to forage properly, to hunt for prey or to escape from predators. For instance, in the case of the cheetah, even slight injuries could prove disastrous and fatal.

In contrast, in captivity, animals are drawn into the orbit of human culture of which veterinary medicine is an aspect. The (human) concepts of diagnosis and treatment of illness, with their elaborate structure of health professionals, are extended to such animals for at least two reasons. First, practical/economic considerations make it imperative that animals in zoos be kept healthy; in general, given the change in climate, the diet, the habitat, and therefore, lifestyle, they are very vulnerable health-wise. To replace them on a regular basis as stock succumbs to disease and death could be expensive. Second, moral/welfare considerations dictate that zoo keepers and managers keep the animals they control as free from pain as possible. In other words, such animals (like domesticated ones) are 'humanised'; contemporary (human) culture, at least in the West, is based on the fundamental moral axiom that pain is evil and that one has a duty to ameliorate pain whenever and wherever one can do so, and to save a life whenever one can.[3] Thus, through incorporation into human culture, such animals lose their ontological status as naturally occurring beings with independent value in their own existence and their respective modes of existence; they become (through no choice of their own) dependent on humans for their very existence and for their changed mode of existence, as exhibits in zoo collections. Their claws and paws have to be examined on a regular basis, their blood and urine tested to determine whether they are well or unwell, and they have to take pharmaceutical products dished out to them in ways convenient to the task at hand. In other words, they have become (clinical) patients, under the care of registered veterinarians and carers. Some zoos may even have hospitals or sick bays to handle poorly animals.[4]

Suspension of natural evolution

According to neo-Darwinsim which, by and large, is the orthodoxy in the scientific domain of natural evolution, animals and their species have evolved through the mechanism of natural selection.[5] This scientific explanatory framework as well as world-view dispenses with the concept of design, whether divine or human. Natural selection amounts to the following key assertion, namely that animals with certain characteristics which enable them to adapt to their habitat/environment would survive to a reproductive age, mate and leave offspring behind. Those with less favourable traits would either die in infancy without reaching sexual maturity, would on reaching sexual maturity be unable to out-compete their rivals in the mating game, would be infertile even if successful in mating, or would have cubs which are still-born or weak in some aspects, and therefore, themselves die before reaching sexual maturity and in turn leaving posterity. A complex chain of causes and effects would ensure that, on the whole, those which survive to reproduce are well adapted to the environment/habitat within which the species has evolved. In this limited sense, and this limited sense only, they may be said to be the outcome of 'competition' leading to 'the survival of the fittest'; they are not 'the fittest' under all circumstances, but only under those specific ecological/climatic conditions to which their species have become adapted over geological time. In the history of Earth, there have been several non-anthropogenic extinctions of species when climatic/ecological conditions drastically changed.[6]

What external conditions in the wild weaken animals and what do they die of? Obviously hunger/thirst, extreme unexpected cold/heat, wounds, disease, predation, and the like, adverse conditions operating either singly or together synergistically.[7] Just to give one example of synergistic causation at work – a hungry animal, provided it is healthy in all other ways, may be able, nevertheless, to escape predation, while one which is only slightly wounded but not hungry would also be able to run away from its predator. However a prey animal which is both hungry and even slightly wounded would be unlikely to do so, as its speed would be thus greatly reduced, by a factor which exceeds the mere adding up of the loss of speed incurred in the two disadvantaged conditions acting in isolation.[8] It is within such a complex causal context that natural selection operates.

In zoos, their captive exotic animals are never left hungry or thirsty (given that they do not have to forage or hunt for their food, as nutritionally adequate meals are served up to them daily as part of 'room service'); they are checked and monitored by health professionals daily/regularly for the least sign of dis-ease and/or disease (at least in the better managed zoos which disseminate best practices throughout the zoo world). Very significantly, they enjoy sheltered accommodation in such a way as to remove the problem of predation altogether. In sum, they are not allowed to go hungry and certainly not to die from famine/thirst, to be (physically) unwell, to die from wounds or disease (in so far as medical science and technology could ensure), to be killed by other animals for food, or through aggression whether in the mating context or otherwise. It is not a surprise, therefore, to learn that zoos pride themselves on providing a more painless as well as longer existence for its animals than their so-called counterparts in the wild.[9]

In the wild, the cheetah lives eight to ten years; in zoos, it could live up to 17 years, almost double their natural life-span. Not only is the wild cheetah cub vulnerable to lions and hyenas, the adult is preyed on by lions and leopards, predators bigger than themselves. In the Serengeti, 90 per cent of cubs die before they are 3 months old. Mothers sometimes have to leave their cubs for over 48 hours as they have to hunt everyday, and should they be unsuccessful in their hunt to build up sufficient milk for their young, they would abandon the litter and start all over again. As for the fate of adult cheetahs, lions in national parks kill as many as 50 per cent of them.[10] No definitive data exist regarding the longevity of (mountain) gorillas in the wild. However, the longest lived gorilla in captivity reached 35 years old; no gorilla in the wild has been known to look as old as that. So it is speculated that in the wild, it may live to between 25 to 30 years of age.[11] In the wild, six out of ten polar bear cubs die in the first year of their existence, either through starvation, accidents or attacks; sometimes, mothers even kill their cubs because they themselves are malnourished, and therefore have to cannibalise their offspring. Sometimes, cubs are prey to adult male polar bears, not to mention other predators such as the wolf. The male also sometimes kills its rivals in the competition for mates, and occasionally even kills females protecting their cubs. However, in general, the adult polar bear in the wild has no natural predators; the successful adult can live up to 25 years or more. On the other hand, the oldest known polar bear in zoos lived to 41 years.[12]

In the three instances of the gorilla, the cheetah and the polar bear, the statistics bear out the fact that such animals under captivity live longer than their wild relatives. However, in the case of the elephant, this appears not to be so, at least according to Clubb and Mason (2002) whose study shows that in the wild, African elephants live up to 65 years; Asian elephants in camps live just as long while individuals who live up to 79 years are not unknown. In zoos, their review of the data from the studbooks 'reveal that out of 517 Asian and 238 African zoo elephants of a known age (dead and alive), none have lived to 60 years, and the maximum recorded age is 56 in Asians and 50 in Africans' (ibid., ch. 7). The average span of survival of elephants in European zoos is 15 years while that of timber elephants is 30 years.

However, Clubb and Mason's study in this matter has since been challenged on the grounds that it is methodologically flawed and that in reality captive elephants in professionally run zoos live just as long as wild elephants – see Weisse and Willis (2004).

From the perspective of this book, the dispute just cited above is of interest for the reason that both sets of disputants, in spite of the obvious difference in their assessments, are, however, agreed on one thing, namely that it matters whether captive elephants live as long as elephants in the wild. Clubb and Mason, based on their unfavourable finding, argue that it may no longer be justified to keep them in zoos, while their critics argue that it is justified to keep them because captive elephants, after all, as a matter of fact do live as long as, even if not longer, as is the case of other animals, than their wild relatives. Both sides, therefore, agree that longevity and indeed (preferably) improving on longevity in the lives of captive animals is a justification for keeping them in zoos. Both sides appeal to the value that the prolongation of life is itself an unquestioned as well as an undiluted good. In human culture, this is no doubt the view held by a majority of people; hence it is generally considered to be the morally right thing to do for the medical profession to strive to save and to prolong life for as long and as far as its skills and technology permit.[13]

However, both sides have uncritically extrapolated this assumption from human culture to the existence of animals as a whole. As this book, so far, has shown, captive animals in zoos are indeed being 'humanised', having been incorporated into the human world of control and management together with the values which underpin such an arrangement; it would follow from such a presupposition that improving the longevity of captive animals is indeed a worthwhile and desirable goal. However, to claim that captive animals living longer lives than analogous animals-in-the-wild constitutes a relevant difference which justifies zoos, thereby also implying that wild animals are suffering unnecessarily, is philosophically flawed, as it ignores the ontological difference in status between animals-in-the-wild that are naturally occurring beings and zoo animals that are 'humanised' through human control and management as exhibits. Zoos, for their animal inmates, are what may be called 'totalising' institutions, in the way long-term prisons and long-stay hospitals/homes are for their human inmates. As freedom of movement is no longer available, naturally their every need (especially physical ones) has to be provided for, the most basic being shelter, food, medication. This amounts to saying that natural evolution working through the mechanism of natural selection as the most crucial factor determining natural evolution in any habitat has been deliberately removed. Death, arising from lack of nutrition, hunger, accidents, disease and predation no longer or hardly obtains. Indeed, as their animals live that much longer, the turn-over is that much lower; as a result, zoos often find themselves in the embarrassing position of over-producing and having to resort surreptitiously to putting down surplus stock. In contrast, death through hunger, malnutrition, accidents, disease and predation is built into the very nature of existence in the wild. However, the context of natural evolution in the wild operating through the mechanism of natural selection is replaced, under captivity, by one of human design and control where natural evolution has been suspended, even though the mechanism of natural selection itself may still be at work, as we have earlier already remarked upon. In nature, death, pain or suffering in itself is neither a good thing nor a bad thing; nature is not capable of making moral judgment. Such matters and events are simply integral to the processes of natural evolution.

As animals-in-the-wild and captive animals belong to different onto-logical categories of being, it makes no sense to claim that one can com-pare them along a commensurate dimension, namely longevity, and conclude thereby that captive animals are better off than wild ones as they live longer. To say that some being that lives longer is better off than another with a shorter existence is only intelligible if both beings are part of (human) culture. It follows that it makes sense to say that gorillas in Zoo A are better off than gorillas in another zoo because they live a few years longer than their counterparts in Zoo B, just as it makes sense to say that people in Japan are better off as they live longer than their counterparts in the United States or sub-Saharan Africa. However, it makes no sense to imply that captive animals are better off as they live longer than animals-in-the-wild. The characteristic of living longer may be a desirable one in the humanised context of zoo animals but is irrel-evant to the context of animals-in-the-wild. An analogy may make this point more clear – as a cat is not a dog, it is pointless to point out that it cannot bark like a dog or that a dog cannot meow like a cat. The ability to bark/yelp is integral to the identity of being a dog (unless that feature is excised via genetic modification), and the ability to meow is integral to the identity of being a cat. For animals-in-the-wild, to die in the way they do, at the age they do, is integral to their identity as naturally occurring beings whose coming into existence, whose continuing to exist and whose going out of existence owes nothing (in principle) to humankind – they are simply the outcome of the processes of natural evolution, which enable them to exist 'by themselves' as well as 'for themselves'. In contrast, captive zoo animals owe their identity to human design and manipulation – they are what they are and live for as long as they do because they owe their very original existence (except for those caught directly from the wild), their continuing to exist and hence, too, their going out of eventual existence (to be determined ultimately by advances in veterinary science and technology as well as by cost) to human control. Zoo animals, as we shall argue in the next chapter, are domesticated (though not in the classical sense of domestication) or immurated animals. It is this profound ontological difference between animals-in-the-wild and zoo animals which render the term 'wild animals in captivity' or 'wild animals in zoos' an oxymoron, a conceptual and ontological mistake which even the more sensitive of professionals and experts of zoos make, alas, all the time.[14]

Conclusion

If the line of argument pursued here and in the last two chapters is plausible, then the book will have made a convincing case so far for maintaining that the ontological status of wild animals is totally different from that of captive zoo animals. The difference may be traced in summary form to the fact that while the former is the outcome of the processes of natural evolution via the mechanism of natural selection, the latter is the product of human design, control and management.

7
Domestication and Immuration

This chapter will examine the claim usually made that captive zoo animals are not domesticated animals or domesticants. It will critically explore the notion of domestication to determine whether or not there is a sense of domestication which may apply to zoo animals. If such a sense exists and can be identified, then this would strengthen the view already arrived at in the last three chapters that, unlike animals-in-the-wild which are the outcome of the processes of natural evolution, zoo animals are the products of human design, management and control.[1]

Domestication: how it is normally defined

Let us first approach this issue by briefly raising the distinction between natural selection and artificial selection. We have already referred to the former in the last chapter; as we have seen, it is an explanatory mechanism accounting for natural evolution, which dispenses with the notion of design, whether divine or human.[2] In other words, (human) intention, whether direct or indirect/oblique, is irrelevant. In contrast, in artificial selection it is the human breeder who deliberately selects a trait or traits of a plant or animal deemed to be desirable with the aim of enhancing and improving upon such properties, leading eventually to the generation of new breeds; conversely, if a trait is deemed to be undesirable, the breeder will similarly try to breed it out. These techniques of artificial breeding have been used for millennia (and are still used in some parts of the world) since the dawn of agriculture and husbandry, until the second agricultural revolution in the 1930s began using the technology of double-cross hybridisation induced by the fundamental discovery of classical genetics by Mendel, the Mendelian laws having been rediscovered at the turn of the twentieth century. From the 1970s onwards, one could say that a third agricultural revolution has arrived in the form of biotechnology based on the fundamental discovery of the structure of DNA by Francis Crick and James Watson in 1957.[3] All three agricultural revolutions resting, first, on craft-based technology, then on Mendelian science and the technology it generates and, third, on molecular genetics as well as molecular biology and the technology they in turn have generated in the last 30 years or so, enabling humankind to select, at a deeper and deeper level of manipulation, genetic characteristics and material containing what we, humans, consider to be desirable traits in a plant/animal and to exclude what we consider to be undesirable traits.[4]

As the above brief account of the history of agriculture shows, breeding of plants and animals is intrinsically tied up with the concept and practice of artificial selection which long preceded the concept of natural selection itself. Indeed, it is said that the phenomenon of artificial breeding and selection inspired Darwin to see an analogue between it and observations he had made in the wild, leading him to formulate the notion of natural selection itself.[5] However, the crucial conceptual difference between the two remains – natural selection dispenses with the notion of design altogether. This conceptual difference is bound up in turn with the ontological difference between the outcomes of natural selection in the context of natural evolution on the one hand, and those of artificial selection on the other – the processes of the former involve naturally occurring beings, while the procedures of the latter involve beings who embody human design and intentionality. The traditional cow who grazed the meadows before the arrival of Mendelian science/technology, the Mendelian cow, bred using the technology of double-cross hybridisation as well as the transgenic cow (who, for instance, produces a human hormone in her milk) are all the results of artificial selection; to labour an important point, they are the products of procedures designed by humans with a particular specific end/outcome in mind.[6]

Having clarified the distinction between natural and artificial selection, let us move on to examine the notion of domestication. Let us begin by looking at an example, the case of elephants. We know that there is a long history of the use of elephants as haulage animals in Asia; from this, some authors have concluded that timber elephants are domesticated or at least semi-domesticated. What about zoo elephants? Are they domesticated or are they wild in all senses of the term 'wild'? We know that they are tame; however, as Chapter 3 has shown, being tame is a necessary but not a sufficient condition for the transformation of an animal to a domesticant. One must, also, straightaway point out that elephants, on any scale, have been kept in zoos for only perhaps a tiny fraction of the time that elephants have been tamed and trained as beasts of burdens.[7] So, clearly, the difference in time-scale would make a difference to the respective outcomes – in the evolutionary history of organisms, the time dimension is crucial, as pointed out in Chapter 1.[8] In the discussion which follows, one is not so much talking about whether zoo animals, as they stand today, are fully domesticated or semi-domesticated in the way dogs are clearly domesticants after thousands of years of breeding, but whether the procedures put in place by zoo keepers/managers, which transform animals caught in the wild and transported to zoos, as well as the majority of zoo animals bred in captivity but whose ancestors were caught in the wild, to become exotic exhibits, as well as the processes of adaptation, which arise from such drastically changed circumstances, have the effect of domesticating such animals, and if so, in what sense of the term 'domestication'.[9]

The concept of domestication has been defined in numerous ways.[10] Here is one definition:

> that process by which a population of animals becomes adapted to man and to the captive environment by genetic changes occurring over generations and environmentally induced development events occurring during each generation.
>
> (Price, 1984 as cited by Clubb and Mason, 2002, ch. 3)[11]

An alternative definition of a domesticated animal or domesticant is as follows:

> one that has been bred in captivity for the purposes of economic profit to a human community that maintains total control over its breeding, organization of territory, and food supply.
>
> (Clutton-Brock, 1999, p. 32)

An influential definition is given by Bökönyi who claims that his is close to that of Clutton-Brock:

> The essence of domestication is the capture and taming by man of animals of a species with particular behavioural characteristics, their removal from their natural living area and breeding community, and their maintenance under controlled breeding conditions for mutual benefits.
>
> (1989, p. 22)

Yet another may be found in Issac in terms of a list of characteristics which animals must possess if they are to count fully as domesticants. Animals which meet some but not all of the conditions are semi-domestic. These conditions are:

1. The animal is valued and there are clear purposes for which it is kept.
2. The animal's breeding is subject to human control.
3. The animal's survival depends, whether voluntary or not, upon man.
4. The animal's behaviour (i.e., psychology) is changed in domestication.
5. Morphological characteristics have appeared in the individuals of the domestic species which occur rarely if at all in the wild.

(1970, p. 20)

The last three accounts of domestication are broadly similar; however, Bökönyi's account holds that domestication is for the benefit of both humans and domesticants – this claim may be too strong, although mutual benefits do happen sometimes.[12] It may have been inspired by the fact that domestication does involve a complex symbiotic human–animal relationship; however, the relationship is, surely, heavily weighted in favour of the domesticator and not the animal to be domesticated.[13]

Clutton-Brock's account in terms of economic profits appears unusually restrictive, ignoring one very important non-economic motive in the history of domestication, especially in its early period, namely the religious motive.[14] Isaac's Condition 1 listed above covers both economic and non-economic motives, and therefore, is preferable to Clutton-Brock's formulation. On the whole, Isaac's account is the best articulated of the four cited; the remainder of this discussion on domestication will bear it in mind, but, however, with two small caveats.

First, it does not refer to genetic changes over generations which Price's definition does. What then could be the relationship between morphological changes (Isaac's Condition 5) and genetic changes? One could say that the latter cause the former, that is to say, all morphological (or phenotypical) changes are caused by genotypical ones; and/or that in turn, all genotypical ones are the result of mutations.[15] However, this way of looking at the relationship is open to two criticisms. First, empirical evidence is usually insufficient in historic cases (based in the main on archaeological sites) to determine that mutations were definitely the cause. Take a slightly different case, of colour variation in modern domestication, cited by Bottema, in which he argues that colour variations in captivity are 'more likely to be the result of recessive factors already present in the wild population than to mutations' (1989, p. 41).[16] Second, the explanation in terms of genetic changes in general ignores environmental causes of morphological changes. For instance, Bottema says that from the unusually large size of the udders of the modern dairy cow, one may be tempted to argue that the morphological change is caused by a genetic change but one would be wrong: 'Although the shape of the udder is hereditary, excessive size is induced by the milking regime, whether milking is done by hand or by machine' (1989, p. 31).[17] Genetic changes cannot be said to account for all the phenotypical changes observed although such changes which are brought about by domestication may be the result of mutations and/or of permitting the emergence of recessive factors under the peculiar circumstances of domestication.

Second, the reference to morphological changes may turn out to be too restrictive; it embraces primarily features of so-called infantilism usually found in domesticants. Domestication involves more than such changes; other biological phenotypical changes also occur at the level of biochemistry/physiology as well as in matters of sexuality, fertility and reproduction in general.

We have so far looked briefly at the concept of what might be called classical domestication. However, none of the definitions quoted above make any explicit reference to the notion of artificial selection.[18] What is the relation then between the two concepts, that is, of domestication and artificial selection? By clarifying the relationship between them, is it possible to identify and articulate another account of what domestication might amount to? Let us see. The notion of artificial selection is, indeed, intimately linked to the classical definition of domestication. Without the concept and the practice of artificial selection, there would be no domesticants in the classical sense. But is artificial selection pertinent only to the creation of domesticants in the classical sense? Can its use in the zoo context lead eventually to animals which can be said to be domesticated in another sense of that term?

The classical type of domestication involves, as we have seen, the deliberate selection of a trait deemed to be desirable or useful in some sense to us, the human selector. If we fancy very small dogs, we breed small dogs with fellow small dogs and after numerous generations of breeding, we end up with the Chihuahua, a dog so small that female owners have been known to pop them in their bosoms. It may be true that some zoos have also indulged in similar practices with regard to some of its animals. For example, an albino tiger may be born in a zoo. Tigers in the wild are rarely born albinos. However, zoo visitors appear to love to see freaks. Such a tiger is a crowd puller and some zoos have been known to give in to such market demands and have tried to breed albino tigers, although zoos with so-called 'best practices' would condemn such a project, on the grounds not merely that it gives in to vulgar clamour but also that it distorts the reality about tigers in the wild.[19]

Artificial selection in the context of classical domestication involves human intention which is direct/explicit, rather than indirect/oblique. Attempts to breed albino tigers – to bring about biological changes – in order to ensure that zoos would always be supplied with such freaks of nature clearly embody this kind of human intentionality. However, today's reputable professionally run zoos claim they do not engage in such scientifically disreputable practices, as we have just remarked. On the contrary, for instance, they claim that individual captive tigers (or whatever other captive animals are said to be endangered) form part of the entire population of zoo tigers, which may be spread out among several zoos throughout the world, and which may participate in their *ex situ* programme of conserving the species (*Panthera tigris*); zoo scientists do their best to replicate the populations of tigers in the wild as far as their genetic variability is concerned. Such zoos deliberately avoid artificial selection of the kind engaged in by less reputable ones when they breed freaks of nature. They imply that, therefore, they engage in no deliberate selection, and also that their populations of captive-bred tigers are not domesticants, that is, in the classical sense of the term.

However, the situation is much more complex than that. In one obvious sense, zoos which participate in specific *ex situ* conservation programmes do deliberately select, say, the individual golden tamarin for the purpose of reproduction, in order to achieve their stated goal of preserving as much, as it is possible, of the species's genetic variability found in wild populations of golden tamarins, in the population of the captive animals in zoos. If golden tamarins in zoos are allowed to reproduce on a random basis of availability and access of animals, then the captive-bred population could, in some instances, reflect the preponderance of the genetic contribution of a particularly successful male, and hence dilute the genetic variability of the population as a whole. Zoos have to intervene to prevent such an undesired outcome. A particular zoo might have to import genetic material from another zoo, which sends the animal or its sperm in order to impregnate its female(s), as it itself has no suitable male for the purpose in hand. Such deliberate selection does not lead to a deviation in genetic variability, as matters stand. For instance, the captive-bred golden tamarins, destined for introduction back to the wild, will, as we shall see in Chapter 9, have been treated in a very different way from other ordinary zoo captive-bred animals not included in the *ex situ* conservation programmes. Scientists, as we shall see, do their best to ensure that these exclusive animals are as little exposed to the processes and procedures of the normal zoo environment as the zoo context of management and control would permit. *Ex situ* conservation zoo animals fail, therefore, to satisfy the classical concept of domestication, in spite of the deliberate intention to maintain the same genetic variability as exists in a wild population, because no particular biological traits (and any underpinning genetic properties) have been directly selected or induced. It, therefore, fails Condition 5 of Isaac's list, as well as Condition 4, which states that 'the animal's behaviour (i.e., psychology) is changed'; as we shall see in Chapter 9, scientists have to take special measures to keep, as much as it is possible, the young captive-bred animals from human sight. As Chapter 9 will also argue, the goal and procedures of *ex situ* conservation do not sit well both in theory and practice with the conception of zoos as collections of animal exhibits open to the public. One may conclude that in spite of meeting the first three of Isaac's conditions, the failure to meet the other two in the special context of *ex situ* conservation, which does its best to minimise the effects of Conditions 1–3, renders that kind of conservation programme outside the confines of classical domestication.

To see if one can find a true departure from classical domestication, one should consider the vast majority of zoo animals, that is to say, those not taking part in *ex situ* conservation programmes and whose *raison d'être* is simply to be part of the zoo's permanent collection of exhibits open to the public.[20] We need to explore in this context a more complex idea based on the relationship between what is directly and what may be said to be indirectly (or obliquely) intended rather than confining oneself to the straightforward version of directly intended artificial selection which we have examined in classical domestication. One also needs, straightaway, to introduce a new concept, that of immuration.

Immuration

This concept may be readily elucidated in terms of some of the highlights in the findings of the chapters preceding this. We list the following:

1. All zoo animals must be tamed, if they are to play the role of being exhibits in a zoo collection. Unlike an animal in the wild, a tame animal permits humans to approach them without showing flight. This is a pre-condition for being a zoo inmate; the same holds true of

a household/farm domesticated (in the classical sense of the term) animal. 'Tame' is the antonym of 'wild' (in one sense of 'wild'). Taming constitutes the minimal content of training which consists of getting the animal used to the numerous ways of being handled by humans. In other words, it is part of a procedure of (human) acculturation.[21]

2. Zoo animals are, in the main, exotic animals. If they are freshly captured from the wild, they suffer geographical/climatic dislocation; and even if they are captive-bred and have known no other environment than zoos, nevertheless it remains the case that they now live in regions of the world which fall way outside the natural home range of their wild forebears.

3. Zoo animals suffer from habitat dislocation as they live in a totally man-made environment. At best, their exhibit enclosures are simulated naturalistic habitats. The operative words are 'simulated naturalistic' as they are designed primarily (though, perhaps not exclusively) to show them off at their best as exhibits. The space permitted them is miniaturised space, which bears no comparison to the size of the home range in which their wild relatives/forebears roam.

4. The loss of freedom to roam is compensated by hotelification. (However, this wrongly presupposes that the need to roam is entirely a means in order to forage/hunt for food, and that it is not intrinsic to the animal in the wild.) Full board and lodging are provided. However, instead of foods found in nature, foraged or hunted, they are given zoo meals which may be nutritionally adequate as diet, but bear no resemblance in all other significant aspects to their respective foods in the wild.

5. Medicare is available and its provision mandatory. They are checked daily for any dis-ease and disease. Should accidents happen and they get injured, they are put right immediately.

6. They are not exposed to the hazards of weather, to the perils of starvation arising from hunger/famine/thirst, to those of wounds/disease which could lead to death. More crucially, they are immune to predation, to being killed in general by other animals. In other words, as we have argued earlier, in zoos, *the forces of natural selection in the context of natural evolution in the wild are in abeyance.*[22] Humans, not nature, are in charge. Under such sheltered accommodation, it is not surprising that they live longer.

7. They are not allowed to choose with whom they mate in three aspects: (a) necessarily they cannot choose their mates from their counterparts in the wild, (b) neither do they necessarily have the choice amongst

conspecifics in the same zoo, as reproduction strategies are under the control of zoo management, and (c) in the case of male animals, they are not allowed to fight for the female they would like to mate with.

8. The features listed above render zoo animals different in ontological status from those of animals-in-the-wild. The latter are naturally occurring beings whose existence is (in principle) independent of human design and manipulation, while in the case of the latter, existence in all key aspects is dependent on humans and their goals and purposes.

The features mentioned above which are intrinsic to zoos collectively define the term *immuration* and its procedures. The term coined comes from the Latin word 'mur' which means wall. A zoological park is not a cage, it is true. The enclosure allocated to a group of animals is undoubtedly many times bigger than a cage; however, the fact remains that an enclosure has walls in the literal sense or limits/boundaries which act as walls (such as a deep ditch around the enclosure). The confines within which movement alone is permitted are designed into such enclosures – zoo animals are captive animals and the space allowed them is allocated by their human managers/keepers.[23]

Zoo professionals might well agree that all the features discussed so far are indeed deliberately designed into the very concept of the zoo. Zoo animals, *ex hypothesi*, have no freedom to roam vast distances, to engage in the lifestyle to which animals-in-the-wild have evolved to do. While admitting this, zoo professionals may well deny that it is part of their direct/explicit intention to create new breeds of animals. However, to say this is to overlook that the notion of direct intention may be understood in ways different from directly intended artificial selection under classical domestication as well as to imply, at the same time, a scientifically misleading understanding of the complex causal relationships between animals and their environments in terms of their adaptation to and their evolution within their new habitats.

We need now to turn to the long delayed task of elucidating the distinction between direct intention and indirect (oblique) intention to see if a more complex version of directly intended artificial selection may not emerge which could cover the case of immurated animals.[24] The distinction is used in everyday discourse not to mention in more specialised ones like legal discourse. The conceptual exploration which follows bears all these discourses in mind:[25]

1. All human acts have consequences which are not merely directly intended, but which can be said to be indirectly/obliquely intended. For instance, drivers directly/explicitly intend to get from A to B using their cars; however, certain consequences follow from their acts of driving; these are congestion and pollution. In other words, these consequences, generally speaking, are unwanted but which, nevertheless, follow from the means chosen to achieve an explicit goal. The goal is directly intended and so is the means chosen to achieve it, as, from the standpoint of a practical agent, whoever desires the end necessarily desires also the means to achieve the end. The car journey as the means to get from A to B is desired and directly intended, but not the congestion and pollution to which the car journey contributes.

2. Consequences other than directly intended ones may be foreseeable or unforeseeable at the time of action. With regard to the former category, one must distinguish between what, as a matter of fact, is foreseen and what is not foreseen at the time of action. In the case of motoring mentioned above, congestion and pollution are both foreseeable and foreseen.

3. Foreseen and unforeseen may in turn be looked at from the perspective of either the individual in question or in terms of the extant collective body of knowledge available at any one time. Congestion and pollution are foreseeable as well as foreseen, both by society in general and by individual motorists (today).

4. Those foreseeable and foreseen consequences, other than directly intended ones, which follow from the explicit goal and the means chosen to achieve it, are often said to be indirectly intended (indeed, sometimes even said to be unintended).

5. Such consequences may be roughly classified as merely probable, highly probable, or amounting to what lawyers call practical certainty.[26] As things stand today, congestion and pollution in the motoring example fall into the last category.

6. On the whole, in the West, the concept of responsibility holds agents responsible only for foreseeable consequences.[27]

7. While there is consensus that agents may be held responsible for a directly intended (bad) consequence of their act(s), there is less consensus whether agents should be held responsible to the same degree for a (bad) consequence which is said in one terminology to be indirectly intended and, in another, even to be unintended. For instance, the law of

homicide in England and Wales wavers between holding a defendant guilty of first-degree murder and holding him/her guilty only of manslaughter, when the defendant claims that s/he did not foresee the bad consequence in question, even though any reasonable person could and would have foreseen it.[28] Consider the case of a defendant who conceived the plan of driving her lover's mistress from town by posting a kerosene-soaked and lit rag through the letter-box of the lover's mistress's house in the dead of night and, as a result, a full-scale fire ensued killing the two children of the lover's mistress, who were sleeping upstairs in the house and were not rescued in time. The former verdict argues that it is not a defence to say that the defendant (being too agitated or too preoccupied with her project) did not, as a matter of fact at the time of posting the lit kerosene-soaked rag through the letter-box, foresee the highly probable consequence of destruction to life; the defendant is, after all, someone with normal intelligence and as a reasonable person, she ought to have foreseen such consequences. Therefore, as a reasonable person, she should also have taken the precaution of ascertaining that the house was empty before posting her burning rag through the letter-box. If her direct intention were simply to frighten her rival in love so that she would leave town, the defendant would and should, as a reasonable person, have taken precisely such measures. As she failed to pursue such precautions, the defence that she did not directly intend to cause death on the grounds that she did not as a matter of fact foresee death is not open to her. The law deems her to have directly intended the death and hence, under this perspective, would find her guilty of first-degree murder, not merely, of manslaughter.

Take another case: the defendant managed to carry out his scheme of getting a bag containing a time bomb into the luggage hold of a certain plane, without anyone knowing that the piece of luggage did not belong to any of the passengers checked in for the flight. The plane in mid-flight would explode at the moment pre-programmed, destroying the plane and killing all the passengers at the same time, and he would then be able to collect a vast amount of insurance money on a policy he had taken out on the plane. He could plead that he only directly intended to enrich himself with the insurance claim regarding the plane, and that he only at best indirectly intended (or even that he did not intend) to kill anyone. No court today, however, would let him off scot free or with the lesser crime of manslaughter and not first-degree murder.[29] This case differs crucially from the burning rag case in one respect – the causal link between the acts (as described by the defendants) and the consequence of death is different. In the burning rag instance, the causal link between posting a kerosene-soaked lit rag through someone's letter-box (the means chosen to drive a rival in love from town) and the ensuing death of two children is not one which amounts to practical certainty, but only to high probability. On the contrary, the causal link in the insurance case between the act of getting a bag containing a time bomb into the hold of a plane (the means chosen to obtain insurance money) and the ensuing destruction of plane and people amounts to practical certainty. In other words, the defendant had done all he could to ensure that death and destruction would occur under the circumstances he had chosen to enact – short of the device failing to trigger off as intended, or the flight being cancelled because of unexpected bad weather, the deaths were bound to occur given the laws of physics and chemistry.[30]

8. Today, jurisprudential thought in the West appears to favour the so-called objective test of liability (the reasonable man test) especially outside the law of homicide; for instance, in environmental law. A firm which pollutes a river cannot simply plead that it is not its direct/explicit intention to pollute the water but simply to get rid of its factory effluent in the most convenient and cheapest way possible; environmental legislation holds such a company liable as it could and should have foreseen (like a reasonable person) that the river would be polluted if it (and others like it) were to use the river in that way. In other words, firms as (rational) agents are held responsible for bad consequences of their acts which are not directly/explicitly intended; the defence is not open to them to plead that at best they are only indirectly intended (or even unintended), as they are readily foreseeable and ought to have been foreseen by any reasonable person. *In this area of law, as in certain types of cases in the law of homicide, the law considers the distinction between direct and so-called indirect intention to be irrelevant, collapsing the latter into the former, especially under circumstances where the causal link between intention and the bad consequences amount to practical certainty.*[31]

The examples of the motorist, the jealous lover, the insurance fraudster and the polluter cited above may also be said to raise the so-called package-deal solution to the problem of intention.[32] The analysis of intention entailed by the package-deal perspective is given in terms of four principles of practical reasoning, in the view, at least, of one philosopher:

Principle of the holistic conclusion of practical reasoning

If I know that my A-ing will result in E, and I seriously consider this fact in my deliberation about whether to A and still go on to conclude in favour of A, then if I am rational my reasoning will have issued in a conclusion in favour of an overall scenario that includes both my A-ing and my bringing about E. . . .

Principle of holistic choice

The holistic conclusion (of practical reasoning) in favor of an overall scenario is a *choice* of that scenario.

The choice-intention principle

If on the basis of practical reasoning I choose to A and to B and to . . . then I intend to A and to B and to . . .

Principle of intention division

If I intend to A and to B and to . . . and I know that A and B are each within my control, then if I am rational I will both intend to A and intend to B.

(Bratman, 1999, pp. 144–5)

However, Bratman has two objectives in mind in formulating the four principles of practical reasoning above: First, he wants to say that those who adhere to the package-deal perspective believe that by applying precisely those principles they would come to the conclusion that the motorist, the jealous lover, the insurance fraudster and the factory owner/manager respectively (directly) intend to cause congestion, to kill the children, to destroy the plane and kill the passengers, to foul up the river. Second, however, Bratman wants to say that the package-deal adherents are wrong to come to that conclusion; they are wrong, he argues, because as his critique shows, one can distinguish between choosing a scenario and intending the expected results which follow upon the choice of that scenario. He wishes, in other words, to maintain that the defendants in the four cases outlined above, as rational agents, could not be said to have (directly) intended the expected results which follow from the choice of the overall scenario. On his reasoning, those results, though expected, are unintentional and thereby implies that the agents who have chosen the scenarios in which such expected results are embedded are not responsible for those results. This seems counter-intuitive. It would be beyond the remit of this book to give either a detailed and full account of Bratman's critique of the package deal perspective or to give a detailed and thorough reply to Bratman's critique. Suffice it here to make the following brief observations regarding the latter:

1. As his account of the four principles stands, they are unable to distinguish between say the insurance fraudster on the one hand and the

jealous lover on the other. We have already observed that the casual link in the two cases differ – in the latter, the link between posting the kerosene-lit rag and the death of the two children as well as the burning down of the house is highly probable, while in the former, the link between planting the time bomb in the hold of the plane and the results of killing the passengers as well as destroying the plane amounts to practical certainty. So one might need to add another principle of practical reasoning:

Principle of causation in terms of practical certainty

> If on the principle of practical reasoning I choose to A and to B when B is not merely highly probable to follow but practically certain to occur, then I intend to A and to B.

2. His problematic distinction between choice of scenario and the denial of intention to do B may be sustainable only because in his fourth principle, he has postulated that the agent knows that A and B are each within his control. The phrase 'each within his control' is questionable especially in cases where the causal relation between A and B is not merely highly probable but practically certain to follow. Bratman cites the example of two bombers to make his case that the distinction between choice of scenario and intention is sound. Terrorist Bomber and Strategic Bomber have both chosen the scenario to bomb a munitions depot, which happens to be next door to a school with children in it, in order to promote military victory. However, only Terrorist Bomber may be said to intend killing the children, although both bombers as a matter of fact would have killed the children by choosing and then executing their choice in action; Strategic Bomber, in Bratman's view, has simply killed the children unintentionally. This is because Terrorist Bomber would behave differently from Strategic Bomber should their chosen scenario in reality start to change when they commence their bombing operations. The Enemy might have got wind of bombs dropping on the munitions depot thereby killing the children, and start to evacuate them. Terrorist Bomber, the Bad Guy, on perceiving the evacuation, nevertheless, would pursue the children and drop separate bombs on them, killing them, whereas Strategic Bomber, the Good Guy, would heave a sigh of relief at the sight of the evacuation and simply concentrate on bombing the munitions depot. The respective reactions of the two bombers to the changing scenario would vindicate attributing intention to kill the children on the part of the Bad Guy while denying such intention on the part of the Good Guy. The scenario

Bratman has chosen to illustrate the difference between the two bombers works for the simple reason that the evacuation of the children which is part of the changing scenario is part of another causal chain, initiated not by the bombers themselves but by the Enemy; furthermore, the foreign causal chain thus initiated, is such that it makes sense for the two bombers to choose to do X (to kill the children in flight) or to do not-X (not to kill the children in flight), something which is within their control.

However, the changing scenario introduced into the discussion has muddied the waters. The two bombers have been chosen especially by Bratman, it appears, to tailor his own conclusion, namely that there is something deeply flawed about the package-deal analysis of intention and that the right intuitive conclusion to draw is that Strategic Bomber does not intend to kill the children, whereas Terrorist Bomber does. Bratman's argument would fail in the case of the insurance fraudster, because while A is within his control (planting the time bomb in the plane), destroying the plane as well as killing the passengers (B and . . .) are not within his control once he has completed putting A into action. B and . . . will follow with practical certainty and there is nothing more he could do, either to hasten the expected results of A or to deflect the expected results of A.

The thought experiment Bratman asked the two bombers to perform is the wrong one to perform. Instead they should have been posed the following question at the time of their deliberation regarding their choice of scenario: Assuming that as expected, the bombs would be dropped and successfully exploded, causing the damage they were expected to cause, namely the destruction of the munitions but also of the schoolchildren, supposing that no divine intervention were to occur in the form of a very strong wind directing the bombs only to drop on the depot but waft them away from the school such that only some but not all the children would be killed or severely maimed, would you in order to achieve the military objective of defeating the Enemy still choose that scenario in question? If the answer to the question is yes, then Strategic Bomber could be said or deemed to intend A and to intend B and . . . . Similarly, the thought experiment posed to the insurance fraudster should be: Suppose that no divine intervention occurred such that the bomb destroyed only the plane but not the passengers as it exploded in mid-air, would you choose the scenario in question? If the answer to that question is yes, then Fraudster intends to A (plant the bomb to get the insurance money) and intends to B and . . . (destroy the plane and kill the passengers).

Posing the hypothetical question in the case of the Fraudster analogous to that which Bratman poses to the two bombers is not helpful: Suppose that miraculously divine intervention occurs to destroy only the plane but spare lives, would you cause another bomb to explode to make sure that the passengers would also be killed? If the answer to that question is no, then Fraudster does not intend to kill the passengers (he merely killed them unintentionally); if the answer is yes, then he does intend to kill the passengers. However, either answer makes no sense as the posing of the question in the first instance is beside the point and absurd: if the bomb (containing the right amount of dynamite calculated to do the job at hand efficiently) successfully explodes in the way planned, then the expected results, which are practically certain to ensue, are not something that are within Fraudster's power of control. Fraudster cannot control or defy the laws of nature. In the real world occupied by real agents acting in accordance with the principles of practical reasoning, no *deus ex machine* appears, no science fiction super-technology exists either; in many contexts, once the rational agent chooses a certain scenario and initiates a certain causal chain of events, the agent has no further control of them since the events unfold according to the known laws of physics, chemistry, biochemistry, biology.

In the light of the clarification above, it may be appropriate to revise Bratman's suite of principles which he attributes to the adherents of the package deal solution to the problem of intention. The recommended revision looks as follows:

Principle of holistic conclusion of practical reasoning
Principle of holistic choice
The choice-intention principle
Principle of causation in terms of practical certainty, and lack of control with regard to the relevant expected results.

The zoo context, of course, has nothing to do with either criminal or civil liability, or even with moral responsibility in general, in any obvious way; however, the revised principles of practical reasoning as set out above are, nevertheless, applicable. It is undeniably the case that human agency is at the heart of immuration which itself constitutes the essential core of zoo activities. The zoo environment/habitat (or scenario in the terminology used by Bratman above) is directly intended and designed for a specific purpose, namely to exhibit animals to the public (A) and such a (cultural) arrangement, in turn, has biological consequences (B) for which an analogous case can be made (analogous to liability in the criminal and civil law described above) for saying that they may be deemed to be directly intended, as they are foreseeable and, indeed, are foreseen to follow (in broad outlines, if not in their minute details) as a matter of practical certainty, given the extant body of scientific knowledge today.[33] A pertinent consequence of keeping animals as exhibits in an entirely human-designed and controlled environment is, in the long run, a new kind of animal, very different in behaviour and biology and, indeed, eventually in genetic variability from those of its ancestor species in the wild.

At this point of the argument, let us go back to Isaac's five conditions and test them against the claim made here that immuration, under the circumstances specified above, counts as a variant of directly intended selection. Condition 1 – the animal is valued and there are clear purposes for which it is kept – is satisfied as zoos are intended to be establishments where animals are kept as permanent collections of exhibits open to the public. Condition 2 – the animal's breeding is subject to human control – is satisfied as the animals have first been isolated and become exotic and, furthermore, are not in general allowed to breed with any other captive conspecifics but in accordance with the zoo's plans on reproduction, which could indeed even include sterilising or controlling their fertility by means of contraceptives. Condition 3 – the animal's survival depends, whether voluntarily or not, upon man – is satisfied as they are kept and fed, looked after in all ways by humans. Condition 4 – the animal's behaviour is changed in domestication – is satisfied as zoo animals are tame, that is to say, they have lost their fear of humans, their flight tendency and flight reaction, not to mention that they have acquired interest (in certain instances) with computers, pie fillings and other human cultural products and artefacts deliberately introduced into the lives of zoo animals. Condition 5 – morphological characteristics have appeared in the individuals of the domestic species which occur rarely if at all in the wild – may be said to be satisfied in spite of certain reservations.

As already argued earlier, Condition 5 should be enlarged to include other important biological changes which have been observed such as earlier sexual maturity, or more intense sexual activities as well as physiological changes which differ from their wild counterparts. Immurated animals are also heavier in weight. Admittedly, these do not constitute morphological changes of the kind normally found in domesticants in the classical sense, such as foreshortened and widened skulls, decrease in size, features which constitute overall infantilism. One should recall that morphological changes of such kinds take time to emerge even under classical domestication; immurated animals have a very short life-history compared to domesticants which are the products of classical domestication. Finally one should also bear in mind that morphological changes are the last to appear in the procedure and process of classical domestication; there is no reason to think that it would be any different in the case of immuration. Bökönyi has put the point well:

> Since domestication is a complex interaction between man and animal, its consequences are influenced by society, economy, ideology, environment, way of life, etc. Any successful definition of domestication must reflect all these possible aspects of the evolutionary process. The result of domestication is the domesticated animal that first culturally and later morphologically differs from its wild form.
>
> (1989, p. 25)

Further support may be found in Price's more recent account of domestication:

> Animal domestication is best viewed as a process, more specifically, the process by which captive animals adapt to man and the environment he provides. Since domestication implies change, it is expected that the phenotype of the domesticated animal will differ from the phenotype of its wild counterparts. Adaptation to the captive environment is achieved through genetic changes occurring over generations and environmental stimulation and experiences during an animal's lifetime .... In this sense, domestication can be viewed as both an evolutionary process and a developmental phenomenon.
>
> (2002, p. 10)

In other words, the cultural conditions are in place for eventual morphological/biological changes and changes in genetic patterns to take place under immuration, changes which are already underway today, permitting a more complex understanding of artificial selection.[34] Artificial selection under classical domestication involves direct selection of specific characteristics, while *artificial selection under immuration involves not so much direct selection of specific characteristics but the conscious choice of an environment (scenario), thereby wilfully permitting certain characteristics to emerge under the cultural Conditions 1–4 put in place by immuration. As the consequences of this kind of artificial selection are perfectly foreseeable and are foreseen with practical certainty to ensue upon the conscious adoption of the said scenario, by biologists in general, and zoo biologists in particular, one can argue that, as a result, zoos may be deemed directly to intend such consequences to happen at present as well as in the near or distant future.*

Such a situation and logic appear to have an analogue in the dim past of humanity, at a time even before classical domestication took place mainly in the Neolithic period, namely when animals were corralled and kept inside stockades. Bökönyi refers briefly to this practice: 'There could be primitive animal-keeping without conscious, but with unintentional, breeding selection . . . (1989, p. 26).' The analogue is far from perfect for the simple reason that a primitive corral and the methods used by our primitive forebears would not lead to an environment as tightly controlled and designed as zoos are today. *Furthermore, given our contemporary sophisticated scientific understanding of animals, one can even go so far as to say that reputable zoos today refrain deliberately and intentionally from selecting for certain characteristics, such as for the white trait in certain animals.*[35] *In other words, paradoxical though it may sound, reputable zoos today directly intend not to undertake artificial selection of traits for the vast majority of their immurated animals in the sense of classical domestication, just as they directly intend to preserve the full range of genetic variability in the small minority of animals which have been chosen for their ex situ conservation programmes and destined eventually to be introduced into the wild. However, it remains the case that they directly intend the zoo environment and all the biological changes to the animals, both in the shorter as well as in the longer terms which confinement within such an environment would inevitably induce.* It is our more profound knowledge of biology which makes it meaningful for us to say that zoos directly intend such outcomes, while exempting our dim and distant forebears from a similar attribution.

However, we shall argue in Chapter 11 that such a policy in so-called reputable zoos could well change, should they become convinced by the plausibility of the arguments mounted in this book. At the moment, they consciously and deliberately refrain from selecting for certain traits (say traits deemed to be attractive by zoo visitors) primarily because zoo experts are of the view that zoo animals are simply 'wild' animals which happen to live in captivity. Zoo theorists and zoo keepers, therefore, deliberately intend and ensure that their exhibits retain the morphological characteristics which their wild counterparts possess in order, first, to sustain the claim that they are 'wild' but happen to be exhibited to the public in zoos, second, to uphold their central justifications of zoos in terms of *ex situ* conservation and education-for-conservation.

However, we have already strenuously argued that the term 'wild animal in captivity' is, conceptually speaking, an oxymoron and, ontologically speaking, it conceals a dissonance between being a wild animal and being an immurated animal or an animal living under captivity in a zoo. We shall argue in Chapter 9 that their two central justifications are fundamentally flawed. If the ontological stance of this book is sound and the critique it mounts of zoos which follows from it is also sound, then even reputable zoos may have to alter their justificatory as well as their policy goals. Chapter 11 shows that should such a reorientation take place, then even reputable zoos may do what they today condemn as unacceptable, that is to say, to go for selection of traits deemed attractive and desirable on the part of zoo visitors, such as colour, and so on. Indeed, they might be even more enterprising than that and endorse the cloning of historically extinct animals, and to create chimeras.

If the main line of reasoning pursued so far is plausible, then one could argue that the procedures of immuration, which are directly and intentionally designed to achieve the end of keeping animals as exhibits in zoo collections, do, nevertheless, have practically certain outcomes which are foreseeable as well as foreseen, and therefore, under the principles of practical reasoning already referred to above, may be said to be directly, rather than indirectly intended (or unintended); such outcomes follow from biological processes at work which can lead to changes in the behaviour as well as the biology in general of zoo animals and their morphology, including changes in genetic variability in the long run, properties which are very different from those exhibited by their ancestors/relatives in the wild. Following this perspective, one could justifiably claim that these procedures and processes constitute domestication in a sense different from the classical account given earlier, but domestication, all the same. E. O. Price, a world-leading authority on domestication, appears to lend support to this view:

the domestication process is difficult to avoid when animals are brought into captivity. Most captive-reared wild animals will express certain aspects of the domestic phenotype simply by being reared in captivity. The application of artificial selection together with the effects of natural selection in captivity can greatly accelerate the domestication process.

(2002, p. ix)

However, this author would not wish to insist on using the term 'domestication' in case it should lead to a futile verbal controversy. As the matter raised goes beyond mere terminology to substantive issues as well as conceptual and ontological matters, it may be best to anticipate a possible red herring and use a different word altogether to characterise the complex issues discussed above, namely, the term *'immuration'*. Zoo animals, in the main, may be said to be immurated animals.

Earlier on, we remarked *en passant* that experts acknowledge that elephants in timber camps may be said to be either semi-domesticated or fully domesticated.[36] Clutton-Brock (1999) argues for the latter. She notes that although elephants in timber camps and camels are not subject to deliberate artificial selection in the way dogs and pigs have been over the millennia, nevertheless they are tamed, trained and exist in a more or less human-controlled environment as an exploited captive.[37] In this aspect, she would be in agreement with the analysis which is emerging, that there is another type of domestication, apart from classical domestication *via* artificial selection of certain traits.

If the environment and lifestyle of timber elephants differ from that of wild elephants, those of zoo elephants differ even more drastically, as we have seen, from those of their wild ancestors/relatives. It is, therefore, not at all surprising that zoo elephants show distinctly different behaviour in several key aspects. For instance, they are much heavier (the result of their zoo diet, offered 'on a plate', and their inability to roam large distances on a daily basis) – the female zoo elephants are 31 per cent to 72 per cent heavier than non-zoo elephants. They engage in sex more often than their counterparts in the wild; the females reach puberty earlier, and could start to breed as early as 11 or 12 years old while their counterparts in the wild or in timber camps reach sexual maturity at 18 years. On the other hand, a third of zoo females fail to breed at all. Furthermore, up to a quarter of calves born to Asian female elephants in captivity are stillborn; up to 18 per cent of the calves may be killed by their mothers (the latter is probably the result of the fact that the family structure of the wild elephant herd is not replicated in zoos). It is reasonable to postulate that such significant differences in behaviour are foreseeable and foreseen consequences of immuration, which are practically certain in broad outlines.[38]

Conclusion

This chapter attempts to establish that classical domestication is not the only kind of domestication at work under artificial selection by arguing that the notion of artificial selection should not be understood simplistically to exemplify only what takes place under classical domestication.

Another version of artificial selection is identified which occurs under immuration, which is just as much a form of domestication. However, in order to avoid potential needless controversies about a futile matter of terminology, this new term has been coined.

Immuration, as well as classical domestication, involves changes in the animals operating at two levels, the biological as well as the cultural levels. In the latter domain, the animals have to be explicitly incorporated into the social structures of human institutions, of control and management, not to mention ownership and exchange. Such animals have to be tamed (to a greater or lesser extent, depending on whether they are wild- or captive-born) and trained to respond to the presence and the peculiar ways and idiosyncratic demands of their human controllers. They have to learn to adapt to another entirely different set of social relationships (which includes humans as their masters/controllers, who also generally impose on them a different family/herd structure regarded as fitting or convenient for zoo policy and management), to different physical environments/habitats, to different lifestyles, including different feeding and reproductive strategies.

These cultural changes in turn set in motion biological processes of adaptation and evolution. In the case of classical domestication, artificial selection occurs for reasons, whether economic, aesthetic or cultural in the broadest sense, leading to the emergence of new breeds. In the case of immurated or zoo animals, a more complex variant of artificial selection than in the context of classical domestication takes place which has two prongs to it. It rests:

(a) On the relationship between the direct intention of immurating animals with the end of exhibiting them to the public (A) and certain results (B) expected to follow causally from (A) which, on further analysis, may be said to constitute direct intention under certain conditions. Zoo experts can foresee and do foresee that immense changes of a biological nature would take place either at present, in the near or distant future once the cultural conditions of immuration are put in place; as these consequences can be foreseen and are foreseen with practical certainty, and as there is no way of preventing them happening short of abandoning the scenario of immurating animals altogether – that is to say, of abolishing zoos as collections of exotic animals for exhibition to the public – the usual distinction between directly and indirectly intended consequences collapses in this context, and zoos may be said to directly intend all the biological consequences which result from immuration. Zoo biologists/managers, who are rational agents acting in

accordance with the principles of practical reasoning, cannot convincingly argue that although they directly intend the means to achieve the goal of exhibiting animals to the public, they only at best indirectly intend all the consequences which follow from the means adopted. (Under Bratman's account of intention, they might even claim, however wrongly, that such consequences are unintended.)

(b) The sophisticated understanding of the biology of animals available today enables zoo experts in so-called reputable zoos to opt deliberately and intentionally not to select for certain traits deemed (for instance, aesthetically) desirable as happens under classical domestication. The omission of such experts to select for, say, albinism is itself a fully conscious act; in that sense they may be said to have directly intended not to endorse artificial selection as it operates in classical domestication; instead, they may be said to directly intend to endorse all the biological changes in the short and long run (B) ensuing from immuration (A) which they directly intend.

Immurated animals, no less than domesticants (in the classical sense) are biotic artefacts, a theme to which we turn in the next chapter.

8
Biotic Artefacts

The preceding six chapters have attempted to demonstrate that zoo animals differ from animals-in-the-wild in profoundly different ways which, in turn, underpin the difference in their respective ontological statuses. Chapter 2 has argued that, ontologically speaking, the latter are naturally occurring beings while the arguments marshalled in Chapters 3, 4, 5, 6 and 7 have as good as shown that the former may be said to be biotic artefacts. This chapter will explore more fully the concept of biotic artefacts as the ontological foil to naturally occurring organisms.

Artefacts

In this context, artefacts refer to human artefacts.[1] Chapter 2 has defined *en passent* an artefact as the material embodiment of human intentionality. Now is the time to elaborate a little on this matter.[2] Another way of making the same point is to elucidate the notion in terms of Aristotle's four causes. Take a statue as the paradigm of an artefact – its material cause is marble, its efficient cause is the sculptor, its formal cause is the blueprint either in the sculptor's head or sketched out on a piece of paper, and its final cause is the purpose for which the statue has been commissioned, such as to commemorate an event or a national/municipal celebrity.

The last three causes refer to human agency and its intentionality; the first to the material medium in which the intentionality becomes embedded. Without human agency and its intentionality, there would only be matter. With the extinction of the human species and its unique type of consciousness, the artefacts which humans have created out of matter would also disappear, leaving only matter behind. The Taj Mahal, as a mausoleum, (which Shah Jahan built in commemoration of his favourite wife, an exquisitely conceived and constructed building made of marble which the world, since its appearance, has come to admire as a great work of art) would no longer exist; only the marble (as a naturally occurring substance), from which it has been made, would continue to exist until the natural actions of wind, rain, plants and animals finally wear it down to soil. The Taj Mahal is an artefact; as such it is a human construct and construction, and, therefore, necessarily it has neither meaning nor existence in the total absence of human consciousness.[3]

Biotic artefacts

While it is evidently clear that artefacts can be made of abiotic matter (like marble) or exbiotic matter (wood), it is not so evidently clear that they can also be made of biotic matter, that humans can create artefacts out of individual organisms.[4] To see how it is conceptually possible to do so, let us further elucidate the notion of artefact, this time, not so much in terms of Aristotle's four causes, but in terms of distinguishing among three theses of teleology, namely, external teleology, intrinsic/immanent teleology and extrinsic/imposed teleology.

We have already observed that Darwin's theory of natural evolution and its mechanism of natural selection dispense with either divine or human design. Such a view amounts to the rejection of the thesis of *external teleology*; the relation between neo-Darwinism and external teleology may be spelt out as follows:

(a) The former denies what the latter asserts, namely that divine agency has created Life on Earth, either in general or in its diverse forms. Neo-Darwinism also denies, for good measure, that human agency has anything to do with the beginning of Life (since Life came into existence and evolved 3.5 billion years ago, as an earlier chapter has already observed), while strictly speaking, external teleology has nothing to say about the matter. It is said that if the age of Earth is made equivalent to a calendar year, *Homo sapiens* only appears 3.5 minutes before the year's end.

(b) External teleology implies that divine agency maintained or sustained Life in its diverse forms since its beginning on Earth while neo-Darwinism claims that neither divine nor human agency is required.

(c) External teleology holds that Life came into being and continues to exist in its diverse forms specifically to sustain human life and to serve human ends, a view which can be traced to Aristotle in his *Politics* and which has profoundly influenced philosophical as well as religious thinking for centuries in the West.[5] Neo-Darwinism denies this claim.

The relation between the thesis of external teleology and that of *intrinsic/ immanent teleology* may be spelt out as follows: the latter may be said to be an implication of denying the thesis of external teleology. In Chapter 2, we have already looked very briefly at it. Individual organisms, as we have seen, exist 'by themselves' as well as 'for themselves'. As autopoietic beings, they strive to keep alive, to reproduce, and so on, but animals-in-the-wild, however, do not do so to fulfill any end or purpose of any external agents; they do so entirely to maintain their own functioning integrity. They do what they do 'for themselves' alone.[6] In appropriating nutrients to sustain itself, the oak produces acorns in order to reproduce itself; it does not produce the acorn in order that the pig may have food. In turn, the pig eats the acorn, simply to maintain itself, and not in order that it would provide a nice dinner for the human hunter. And when the lion eats the human who has just dined off the roast pig, it is just as true to say that the human has not eaten the pig in order that he might himself, in turn, satisfy the lion's hunger. Of course, as a matter of contingency in general, organisms find certain other organisms useful in sustaining their own functioning integrity – plants that are primary producers, nevertheless find insects helpful in propagating their pollen, and certain mammals useful in propagating their seeds.

Organisms, in living 'for themselves', are realising their respective *tele* as individuals and as members of their species.[7] In so doing, they exemplify the notion of intrinsic/immanent teleology. An adult female frog will mate with her male counterpart. Her fertilised embryos will develop into tadpoles; in turn, the tadpoles will grow into adult frogs. As they are frogs, not birds or wolves, they can only live or thrive in certain habitats and not others; they prey upon some other organisms, like insects, but not others. In all ways, they behave as they do, entirely in accordance with their own *tele*, independently of human agency and its manipulation. Their trajectories, both as individuals and as species, have, in principle, nothing to do with humankind. In the absence of humans in the world, they would be there, coming and going, at their own pace and in their own ways. In other words, they follow their own trajectories.

Humans may turn naturally occurring organisms into artefacts, just as much as they may turn abiotic matter into artefacts. Such attempts exemplify the thesis of *extrinsic/imposed teleology*. As we have seen, humans have been domesticating plants and animals for a very long time. For millennia, their success rested on using what this book calls the craft-technology of selective breeding; however, in the first half of the twentieth century these traditional methods, as we have already mentioned, were radically overhauled by a new technology, which was informed by the theoretical understanding given by the science of classical Mendelian genetics. The last quarter of that century also witnessed, as we have seen, the arrival of a yet more powerful technology, called biotechnology or genetic engineering, which is informed by the theoretical understanding given by the even more basic sciences of molecular genetics and molecular biology. It is more powerful precisely because it allows humankind to cross boundaries between species and kingdoms by manipulating organisms, no longer at the level of whole organisms but at the molecular – DNA – level. For instance, one can insert into bacteria, DNA that may belong to the human genome. As we have seen, one can get cows to produce human proteins in their milk.

The examples just mentioned illustrate the process of transforming naturally occurring organisms, as in the case of the bacteria, to become biotic artefacts, or in the case of the cow, which as a domesticated animal is already a biotic artifact, to embody a deeper level of artefacticity.[8] The transgenic cow, unlike the more usual domesticated cow, has been commandeered by humans to use its autopoietic powers of self-maintenance to produce, not cow's milk, but milk which contains a human protein. In other words, biotechnology has succeeded in severing, in the clearest manner possible, what has been an inseparable link between being an organism, which exists 'by itself', and one which exists 'for itself'. Up to even 25 years ago, the distinction between 'by itself' and 'for itself' was one that could only be made intellectually but not empirically. But recently biotechnology has managed to sunder them as a matter of fact.

The transgenic cow, *par excellence*, no longer exemplifies existence 'by itself' – *ex hypothesi*, such an organism would not have existed without the direct and deliberate intervention of humans. The same is true of the transgenic bacteria. Humankind, *via* biotechnology, has captured the biological mechanisms of cows or bacteria in order to make them be what humans want them to be, and not how they themselves would be in the absence of human manipulation and control. In other words, although they may still perform biological functions such as eating, breathing, digesting, nevertheless, in carrying them out, they have been made to subvert their own respective *tele*. The cow no longer produces cow's milk, fit to nourish her own offspring, in principle at least when the milk is not whisked away for human consumption. Instead, she is made to produce milk whose constituents are not those in accordance with her *telos* as a cow. The same is true of the bacteria – their own *tele* have been subverted and made to execute a human intention and human end instead. This embodies the notion of extrinsic/imposed teleology. However, the transgenic organism, made possible by biotechnology, is but the most extreme form, to date, of a biotic artefact.

As we have shown in some detail in Chapter 2, the fact that organisms, in maintaining their own functioning integrity, exist 'for themselves' has not stood in the way of human success in transforming them to become biotic artefacts. Biotic artefacts, as much as abiotic artefacts, are not autonomous, as they are no longer *simpliciter* naturally occurring entities. However, in spite of the profound similarity of sharing the same ontological status of being artefactual entities, there is a residual difference between them. Imagine the sudden disappearance of *Homo sapiens* from the face of Earth. Empirically, over time, abiotic artefacts like houses, jewellery, computers would no longer exist; such artefacts, without constant human maintenance and repair, as we have mentioned earlier, will just disintegrate and eventually become dust and/or soil. But at the philosophical level, in the absence of humans and their type of consciousness, there would necessarily be no human artefacts, just chunks of physical matter, even before they disintegrate and decay. The other sentient beings which remain, like the leopard or even the chimpanzee, would not and could not know these things as human artefacts; only another being with a similar sort of consciousness like ours could recognise them as such.

But in the case of cows or pigs (whether transgenic or the more ordinary non-transgenic ones), in the absence of human maintenance many of them would die, or might not even succeed in reproducing themselves. But some might survive, and after many generations over evolutionary time, the humanly selected characteristics or the inserted transgenic material might become very rare in the genetic makeup. They could become naturally occurring again, like feral pigs, except for their remote genetic history. What this implies simply is that natural evolution and its processes (in the absence of human manipulation and control) have their own trajectories.

As we have argued in the last chapter, immurated animals – just as much as classically domesticated animals and transgenic animals – are the result of artificial selection, the difference lies merely in the fact that immurated animals are the outcome, in the main, of artificial selection within a context which is much less straightforward than its use in the context of classical domestication. Both forms of selection are manifestations of the thesis of extrinsic/imposed teleology which issue in animals whose main *raison d'être* is to serve human ends/purposes, rather than to follow the trajectories of their ancestors in the wild; they no longer live 'by themselves' and they no longer live 'for themselves', as they are biotic artefacts.

Type/token distinction revisited

In Chapter 2, we raised the distinction between type and token in the context of naturally occurring organisms. Here we explore two things: the distinction in the context of immurated animals as biotic artefacts and in the further context of assessing whether, from the ontological standpoint, an immurated animal can be said to be a token of a naturally occurring species in the way its wild ancestor/relative can unproblematically be said to be a token of the species.

But first, a few words about the type/token distinction in general. The actual letters of the Latin alphabet which are used here to write this book are tokens of a type, namely, the letters of the Latin alphabet as opposed to the letters of another alphabet, say, the Greek one. The individual Rolls Royce is a token of the type of car called the Rolls Royce. The word 'type' is usually used in reference to abiotic/exbiotic artefacts, or to abstract constructs such as letters of an alphabet or of musical compositions (the rendering of Beethoven's Ninth Symphony by the BBC Philharmonic Orchestra is a token of the type which once existed in Beethoven's head, in outline at least, if not in all absolute details, but which he had committed to paper using musical notations).

In the context of biological classification, the term 'type' is not usually used; instead, one uses the term 'species'.[9] More precisely, the term 'species' used in this book refers, in the main, to two different, but related understandings, namely, in terms of the biological-species concept as well as the evolutionary-species concept. As Chapter 2 has already touched on the latter in sufficient depth, no more will be said here except to remind the reader that zoos, by rendering animals exotic, have deliberately wrenched them from their evolutionary context and past; that such animals and their descendants have no choice but to live in an environment within which the processes of natural evolution and natural selection have been suspended (as shown in Chapter 6). A few brief words, however, will now be said about the former. Chapter 7 has argued that one must recognise that the human-designed habitat of the zoo does induce profound changes in the animals' behaviour and that immurated animals do adapt to such an environment and evolve within it. Such adaptation and evolution would eventually (given a long enough time-span) lead to even new species, different from the species to which their ancestors/relatives in the wild belong. This is in accordance with scientific understanding and is not at all far-fetched.[10] Furthermore, according to the biological-species concept normally used in biological classification (in spite of some obvious drawbacks), a species is distinct from another if the members of the one do not naturally meet, mate and reproduce with members of the other, even if they can do so biologically.[11] Lions and tigers belong to different species, as they do not as a matter of fact meet, mate and reproduce with each other in the wild owing to the fact that they live in very different geographical locations and habitats. However, if humans were to intervene and mate a lion with a tiger, offspring may issue – those born of a lion father and a tiger mother are called liger, while those born of a tiger father and a lion mother are called tigon. In fact and in practice, zoo animals cannot and do not meet, mate and reproduce with their counterparts in the wild, as they occupy very different geographical locations and habitats. However, should humans wish, they could capture an animal from the wild and make it meet and get it to mate with its counterpart in a zoo; alternatively, under *ex situ* conservation programmes, captive-bred animals may be released in the wild so that they can meet, mate and reproduce with their counterparts in the wild. Neither of these two possibilities is available in reality for the vast majority of zoo animals. They remain captive-born and captive-bred from generation to generation, setting in motion certain biological processes, which would eventually lead to breeds distinct from the species of their wild forebears.

The individual polar bear in the wild is a token of the naturally occurring species *Ursus maritimus*, both in the biological-species and the evolutionary-species senses of the concept of species. It mates with fellow polar bears forming membership of the polar bear population(s) in the wild; it is a descendant of ancestors through long, unbroken, evolutionary processes in natural evolution in their home range, leading back 100 000 to 250 000 years ago when the species first emerged.

On the other hand, the individual immurated polar bear in a zoo does not belong to the species *Ursus maritimus* in the sense understood by the biological-species concept; as we have seen, such individual immurated polar bears in zoos would never as a matter of fact meet and interbreed with individual polar bears-in-the-wild, except in cases of deliberate human intervention. At the same time, such immurated individual polar bears live within a cultural space designed, maintained and controlled by humans in such a way that their very existence is a fundamental rupture from that of polar bears-in-the-wild. While those in the wild are descendants of an unbroken historical evolutionary lineage of *Ursus maritimus*, immurated individuals live their lives in a context within which the processes of natural evolution are suspended. In other words, polar bears-in-the-wild and immurated polar bears belong to two crucially different categories or types, ontologically speaking – the former are naturally occurring beings while the latter are biotic artefacts. Hence, the correct inference to make is that immurated polar bears are not tokens of the naturally occurring species, *Ursus maritimus*, in either of the two senses of the concept identified, namely the biological-species concept and the evolutionary-species concept. To mark the ontological difference between them, one could propose a new system of classification and say that the individual immurated polar bear is a token of a different species to be called *I(Ursus maritimus)*, where *I* stands for 'immurated'.[12]

Conclusion

This chapter pulls together the various strands of arguments presented in the preceding chapters (2–7) which serve to establish the main thesis of this book, namely that from the ontological perspective, animals-in-the-wild and zoo animals belong to two fundamentally different types or kinds; that animals-in-the-wild and the naturally occurring species to which they belong are the ontological foil to immurated animals and their corresponding I-species; that while the former are the outcome of natural evolution, the latter are biotic artefacts, the outcome of artificial selection/breeding. They are respective tokens of their respective types. *The World Zoo Conservation Strategy* (1993), in company with most publications on the subject, is just simply wrong when it claims that 'all zoos exhibit living specimens of wild animal species . . . .'

9
Justifications Deemed Serious

Zoos started off as the private collections of kings and princes, aristocrats and the very rich. Modern zoos began in the eighteenth century, open initially to members only, who were interested in exotic animals from the point of view of scientific research.[1] However, in the nineteenth century they became increasingly municipalised and democratised; civic pride and prestige required that every major city in the industrialising West should have a zoo, open to the public, whether free or for a relatively small entry fee. Zoos were meant in those days to be the 'green lungs' in urban settings, where nature, domesticated, was created with trees, where people could escape from the bustle and the pollution of great cities.

The aims promoted by these original zoological institutions appeared to have been more or less the same as those proclaimed by zoos today, namely, research, captive breeding and recreation. However, unlike nowadays, they do not seem to have emphasised education of the public. Today, zoos prioritise their goals as follows: on the one hand, education of the public, conservation of species and scientific research, and on the other, recreation. Unlike in the nineteenth century, some contemporary zoos appear too embarrassed even to mention recreation as a justification, as it is considered to be, on the whole, an undignified and ignoble goal, best ignored and forgotten. However, this chapter hopes to show that the so-called serious justifications of zoos are in fact deeply flawed; the following chapter will, in turn, argue that perhaps the so-called ignoble one in terms of providing recreation for the masses may, paradoxically, survive critical scrutiny.

Scientific research

After the French Revolution, the Jardin de Plantes in Paris started also to house animals. One of its objectives was to contribute to the advancement of science. In 1792, H. Bernardin de Sainte-Pierre argued that it was not enough, in determining their classification in terms of species and genus, only to study the skins and bones of dead animals, but that one ought also to observe the development of living animals. He also proposed to create in zoos what we today call simulated naturalistic habitats in order to make study of their behaviour all the easier. As his perspective was considered too revolutionary (in spite of the fact that Buffon had earlier advocated it), no notice was taken of his plea until the beginning of the nineteenth century, when Frédéric Cuvier, who became the keeper of the animal collection in 1804, initiated studies of the animals under his charge in terms of their behaviour, intelligence and sociability – the zoo became a kind of laboratory in which the observations made would supersede those which one might undertake in the wild. However, not even Cuvier got very far with his project; he met with stiff criticism from scholars like Etienne Geoffroy Saint-Hilaire, who argued against it on both pragmatic and theoretical grounds: that the high mortality rate of captive animals would prevent long-term observation of their behaviour, that an individual animal could not be said to represent the species, and more significantly (from the standpoint of this book) that captivity distorted behaviour.[2] In other words, we see that a debate had arisen – even at the very beginning of the modern zoo in the early nineteenth century – which continues to be relevant today.

Today's zoos are not preoccupied with the problem of classification in terms of species and genus; instead, their scientific research is based on the study of the behaviour of the animals in their collections, as well as of their physiological, metabolic processes and other related aspects, especially those pertaining to their health and well-being under captivity.[3] However, this is not to deny that it can and does address itself to other issues outside the specific concerns of veterinary science and husbandry, leading to significant discoveries about the animals being studied; for instance, in the case of the elephant, the following discoveries have recently been made by scientific zoo researchers:

Work on zoo animals has . . . advanced our knowledge of elephant communication . . . anatomy . . . and reproductive biology . . . Zoo elephant research has also enabled the development of many techniques likely to aid in *ex situ* conservation, such as DNA extraction from faeces and ivory, new methods for monitoring movement patterns, the assessment of reproductive state from faecal hormone metabolites, and new potential methods of contraception . . . such research opportunities would seem the greatest benefit of keeping elephants in zoos, although arguably they also could be supplied by logging camp and orphanage animals.

(Clubb and Mason, 2002, ch. 3)

While appreciating that such research in zoos may have added to the sum of human knowledge about animals, one should also be aware of certain methodological limitations and flaws inherent in such research. For a start, the scientific research appears to be conducted from the standpoint of functional biology. According to Ernst Mayr:

the functional biologist is concerned with the operation and interaction of structural elements, from molecules up to organs and whole individuals. His question is 'how?' How does something operate, how does it function? The functional biologist attempts to isolate the particular component he studies, and in any given study he deals with a single individual, a single organ, a single cell, or a single part of a cell. He attempts to eliminate, or control, all variables, and he repeats his experiments under constant or varying conditions until he believes he has clarified the function of the element he studies. . . . The chief technique of the functional biologist is the experiment, and its approach is essentially the same as that of the physicist and the chemist. Indeed by isolating the studied phenomenon sufficiently from the complexities of the organism, he may achieve the ideal of a purely physical or chemical experiment. In spite of certain limitations of this method, one must agree with the functional biologist that such a simplified approach is an absolute necessity for achieving his particular objectives. The spectacular success of biochemical and biophysical research justifies this direct, although distinctly simplistic, approach.

(Mayr, 1988, pp. 24–5)

However, in spite of its undoubted strength, such a research orientation pointedly ignores that of evolutionary biology.[4] As we have argued in Chapter 1, the zoological conception of animal presupposes the biology of time and history, of the notion of reciprocal causation in the dynamics of organism-in-the-environment, as well as of the philosophy of ecocentrism. Furthermore, the results of observation and experimentation from the perspective of functional biology conducted within zoos must remain theoretically suspect, as the doubts regarding them spring fundamentally from the fact that the animals under study are captive animals, whose very condition of captivity might have a profound effect upon their behaviour, a criticism mounted by Sainte-Hilaire, as pointed out above, two centuries ago. If so, one should hesitate to extrapolate from the results thus obtained to problems which arise outside of zoos regarding wild animals in their natural habitats. It may be sufficient to cite only two recent examples to illustrate the methodological traps lying in wait. The first concerns subjects which are laboratory rather than zoo animals; however, this difference is not germane in this context. A behavioural scientist in California has published his findings in *Nature*, showing that caged parrots show stereotypic behaviour caused by brain damage (damage to that part of the brain called the basal ganglia).[5] The brain damage, leading to stereotypies, appears to have been caused by sheer boredom in confined space. Some attempts to remove boredom through enrichment programmes are indeed successful; however, the point that this author wishes to make here is somewhat different as it concerns a matter which cannot be remedied by simple enrichment in confined space. It is to do with the confined space itself.[6] We have already seen in an earlier chapter that, in particular, medium to large mammals in zoos are routinely kept within enclosures whose severely reduced dimensions are totally out of proportion to the size of their occupants and the distance which the wild ancestors of their occupants traversed daily within their home range.

Another concern is related to the fact that the captive environment is necessarily much simplified compared with the infinitely richer, ever-changing and complex environment within which animals-in-the-wild operate. The zoo lifestyle, which bears no resemblance to the existence animals-in-the-wild lead, may induce differences in behaviour, as the next example – cited by Steven Rose – confirms. As we have earlier argued, immurated animals lack control of their lives in any meaningful way since they exist essentially as exhibits in a human purpose-designed and managed environment. The conditions under which captive animals are reared, as we have also already observed, have a significant impact on adult behaviour – for instance, captive-born animals are, from the moment they are born, exposed to the presence of humans. In other words, they are essentially born tame. According to Rose:

> [b]ack in the late 1920s, the anatomist Solly Zuckerman reported strong dominance hierarchies and high levels of 'aggression' and fighting among the large but confined hamadryas baboon colony at

London Zoo, and developed an influential theory of social behaviour based on these studies. 'Each baboon', he wrote, 'seemed to live in perpetual fear lest another animal stronger than itself will inhibit its activities.' Violence was a constantly occurring event, quarrelling frequent and widespread, and any major disturbance of the precarious equilibrium caused the social order to collapse into 'an anarchic mob, capable of orgies of wholesale carnage.' Later researchers observing baboon colonies in much larger enclosures or in the wild failed to find similar levels of fighting. Instead, the groups seemed relatively peaceful and stable. It became obvious – and with hindsight it seems scarcely surprising – that the behaviour of Zuckerman's baboon group had been dramatically modified by restricting the space within which its members had to coexist.

The constraints of Zuckerman's reductive approach had transformed the situation he wished to study and fundamentally misled him, even though his observations within that constrained situation were presumably perfectly accurate.

(Rose, 1997, pp. 28–9)

It seems fair, in the light of all the points made above, to conclude that any scientific findings about tame and immurated animals may not be applicable to animals-in-the-wild. Uncritical extrapolation from such data is hazardous, and therefore would be unwise; it would amount to bad science. To be methodologically correct and accurate, observation and study of behaviour of wild animals should be made in the wild, not in zoos.

Conservation

This aim has become dominant even if it has not been promoted to be the number one priority in the list of justifications of zoos. However, judging by some of the key documents issued in the last ten years or so on the matter, one might be forgiven for thinking that the sole *raison d'être* of zoos is conservation, from which it follows that should a zoo fail to meet it, such a zoo ought to be closed down.[7]

Two key documents have been exceptionally influential in setting conservation as a key goal, if not the only goal, of zoos. In 1993, the *World Zoo Conservation Strategy* (*WZCS*) was published by the World Association of Zoos and Aquariums (WAZA): 'For both ideological and practical reasoning, nature conservation must be the central theme of zoos in the future (*WZCS*, 1993, ch. 1)'. In 1999, the European Union

(EU) issued its Zoos Directive, which requires all member states to ensure that their zoos commit themselves to the goal of conservation and to conform to the conservation measures it lays down.[8] The directive is hospitable to *ex situ* as well as *in situ* conservation.[9] However, unfortunately regarding the latter, it is obvious that it is not something that zoos can, under normal circumstances, single-handedly secure, although in reality it may be true that it could contribute in some small ways *via* its research (but bear in mind the methodological limitations of zoo research raised in the preceding section), as *in situ* conservation necessarily requires action on the parts of sovereign states with the pressure, support and encouragement of concerned world bodies, such as the United Nations, to protect whole ecosystems and habitats.[10] In other words, it is expected that zoos are best equipped to contribute to *ex situ* conservation; in particular, *via* the techniques of captive breeding, which zoos themselves proudly claim to have so successfully pioneered.[11]

This book has no desire to rehearse in detail the criticisms usually mounted against the goal of *ex situ* conservation *via* captive breeding, except to remind the reader that: (a) it is very expensive; (b) it can, at best, save only a few species; (c) saving a few species in isolation from saving their habitats is mistaken in theory and in principle; (d) given these points, its critics argue that it is neither cost effective nor sound to engage in *ex situ* conservation, and that resources devoted to it should be diverted to *in situ* conservation instead.[12]

However, this chapter will develop a critique based on the following points which will in the process take up criticism (c), mentioned above, in particular:

1. The definition of zoo.
2. The exact nature of the zoo animal which *ex situ* conservation intends to save from extinction.

Imagine for one moment that the collective efforts of zoos throughout the world in their captive breeding programmes over the next 150 to 200 years were able to save all the animals on today's lists of endangered/threatened animals, and that zoos have become a true Noah's Ark. Would this mean that zoos are justified in claiming that their chief, if not only, goal is conservation? In other words, can zoos even be defined in terms of such a goal? If so, then the definition(s) of zoos as they stand today would have changed. But, how are they defined today?

According to the 1999 *EU Zoos Directive*, zoos are 'permanent establishments where animals of wild species are kept for exhibition to the public for 7 or more days a year'.[13] Section 21 of the Zoo Licensing Act 1982 (UK) says that a zoo is

> an establishment where wild animals are kept for exhibition to the public otherwise than for the purpose of a circus and otherwise than in a pet shop; and this Act applies to any zoo to which members of the public have access, with or without charge for admission on more than seven days in the period of 12 consecutive months.[14]

According to *WZCS*, institutions which call themselves zoos have two features in common:

1. Zoos possess and manage collections that primarily consist of wild (non-domesticated) animals, of one or more species, that are housed so that they are easier to see and to study than in nature.
2. Zoos display at least a portion of this collection to the public for at least a significant part of the year, if not throughout the year.

(*WZCS* (1993, ch. 1.3)

In other words, the three definitions cited above agree that zoos are collections of wild animals for exhibition to the public. They, thereby, differ crucially – judged from the ontological view point – from the definition pursued by this book which stipulates that zoos are collections of animals, which are not wild but immurated, for exhibition to the public.

Note, too, that all three definitions incorporate a reference to what looks at first sight a mere administrative detail, namely, that zoos must open their collections or certain parts of their collections to the public for a limited number of days or for significant parts of the year. However, the opening-time qualification in particular is paying more than attention to administrative niceties, as it permits the *WZCS* to reconcile two things which on the surface appear to conflict, namely that zoos must make their collections of exhibits open to the general public (in pursuit of the goal of educating them as part of the mission of today's zoos, not to mention the more vulgar goal of entertaining zoo visitors) and that at the same time they are also expected, in order to attain the goal of conservation, to pursue captive breeding and related scientific activities as laid down by the current agenda of the aims and objectives of zoos. Yet a moment's reflection would show that while the goal of educating the public or that of entertaining the public *ex hypothesi* renders public their collections of exhibits, the other goal of conservation requires the very opposite, that the animals be kept away not only from the presence of the general public but even of the scientists themselves, that is to say, they must be isolated from human presence as much as it is practically possible, as an earlier chapter has already pointed out. The activities of captive breeding with the aim of reintroduction to the wild are incompatible – practically, conceptually and ontologically – with those arrangements associated with exhibition. The latter requires public access, the former its denial.[15]

Clear thinking, therefore, demands that those scientific activities related to the goal of *ex situ* conservation be detached from zoos and be conducted within private space, which may officially be designated as (*ex situ*) conservation research centres, perhaps. The proponents of the WZCS might well retort that one might choose to define terms however one wishes, *à la* Humpty Dumpty, and that critics, like this author, are being unnecessarily prescriptive in confining the term 'zoo' only to its normally perceived function of exhibiting their collections of exotic animals to the public. However, two considerations are pertinent to restraining this kind of definitional free for all: first, differences of verbal matters should be distinguished from those of substantive matters, and second, there are very important issues of substance involved in this particular context, namely that the activities, as currently carried out by zoos under exhortation to achieve the goal of *ex situ* conservation in particular via captive breeding and preparation for introduction to the wild, conflict with those associated with the requirement of exhibition on the part of zoos. It may be true that definitions of terms, *per se*, are merely lexicographical and verbal in character; however, the choice of definitions should be guided by other relevant considerations such as the epistemological value of clarity and the ontological value of recognising fundamentally different categories of beings that the domain of activities deals with.

In light of the above, one could even be provocative and hold that a zoo is no longer a zoo when it engages in scientific research involved in the pursuit of conservation, of the *ex situ* kind, which relies, in the main, on the techniques of captive breeding. To put very different activities – different from the ontological standpoint – under one definitional roof, so to speak, would profoundly mislead and confuse not merely the lay public but also even the scientists themselves. In other words, both the *WZCS* and the *EU Zoos Directive* are fundamentally wrong-headed and mistaken in promoting (*ex situ*) conservation as the central goal of zoos; the remit of zoos cannot be stretched quite so far as to include such activities.

This fundamental misapprehension will become more obvious as we explore the next issue, namely the exact nature of the exotic animals kept in zoos which scientists are expected to save from extinction.

We have already argued in earlier chapters that animals in captivity live under conditions which are so profoundly different from those in which animals live in the wild that the individual animal belonging to the former category cannot be said ontologically to be a token of the species to which the latter individual animals belong. As animals kept in zoos are biotic artefacts living in an environment designed, controlled and managed by humans for humans, where the processes of natural evolution are suspended, they cannot be said to be wild in any sense of that term. For a start, to labour a point already tirelessly made, it is very wrong-headed, both conceptually and ontologically speaking, for the three definitions of zoos cited above to refer to zoos as places which house 'wild animals', or 'animals belonging to wild species'. This book argues that zoos are places which nurture immurated animals, an ontological foil to wild animals.[16]

We have also seen that in one sense scientists and other related professionals working in captive breeding programmes appear aware of the fact that if captive-born and captive-bred animals are to be returned to the wild, such animals must be subjected to a different regime from those captive animals which are to remain permanent residents of zoos to be transformed into becoming full-blown immurated animals. These specially selected animals must, from the moment of their birth, be protected from exposure to humans as much as possible. As we have already seen in the last chapter, zoos are very careful indeed to ensure that as many procedures of immuration as possible are suspended in the reproduction as well as in the rearing of the captive-born.[17] For instance, we know that imprinting occurs in some animals. In the captive breeding in California of condors for return to the wild, the scientists involved did not show their faces or themselves, and their hands were covered by puppet gloves in the shape of adult condors when handling the young birds. This and other similar practices are an acknowledgement by the scientific authorities that the procedures of immuration can and do have crucial impact on the behaviour of zoo animals which would render them different from those displayed by their ancestors/relatives in the wild.[18] They also appear to be aware of another fundamental fact, namely that successful reintroduction to the wild requires that the captive-born and -bred animals be initiated into the culture, so to speak, of the species to which their wild ancestors belong.[19] In other words, some conservation scientists have learned to appreciate that the animals in their charge cannot be expected to survive, to reproduce, to lead successful lives in general in the wild, if they were only exposed to the 'room service' facilities of hotelification laid on by zoos. So they do their best to make their charges acquire, as much as feasible, the skills of living in the wild before letting them go into the unknown future in an unfamiliar habitat.[20] They seem to perceive that an animal cannot be fully understood except in the context of the environment in which it is born, in which it grows up while at the same time learning how to negotiate it not simply by personal individual trial and error, but in the company of its elders, its relatives which show it the way.

Yet the grasping of this truth – that a habitat in the wild nurtures and produces a wild animal while, on the contrary, an environment which is designed, controlled and managed by humans nurtures and produces a biotic artefact – is but dim and intermittent. This charge of partial understanding may be illustrated by one example of success much acclaimed in the literature of conservation, namely the saving from near extinction of the Père David deer (*Elapharus davidianus*), although it still remains classified today as a critically endangered species. The history of this animal is pertinent to the point made here; so some details will be given. Some accounts say that the deer had become extinct in the wild about 2000 years ago and others, at least 1000 years ago.[21] It survived by the nineteenth century only in parks, in particular, the Imperial Hunting Park in Peking (Beijing). A French missionary – Père David – came to know of its existence there, and persuaded the Emperor to allow some animals to be sent to Europe. Shortly after this, in May 1865, catastrophic floods breached the walls of the Park, and starving peasants raided and killed most of the animals for food. Those few that survived were killed a little later during the Boxer Revolution. However, by pure chance, the Duke of Bedford procured a few of the arrivals in Europe for Woburn Park; these bred. More than a century later, in 1986, 39 individuals were introduced in the Dafeng Reserve in China. The species's original habitat is said to be swampy, reed-covered marshlands; the Dafeng Reserve is a coastal, seasonally flooded area.

It is interesting as well as significant to emphasise that very little is known of its original habitat and that the animal has been extinct in the wild for roughly 1500 years, if not more. Furthermore, those individual animals introduced to China since 1986 are the descendants of animals which themselves have become twice exotic (in the technical sense), residing in a park in Britain for more than 100 years. An animal with such a history could not be said to be wild in any sense of the word. Whether the park was an imperial hunting park in Peking, or whether it was Woburn in England, the environment remained a park. A royal hunting park is a human-managed and human-controlled environment in which some of the deer under its care were destined to be killed by the royal hunters, unlike Woburn Park, where the feature of hunting was indeed absent. In other words, the deer cannot be said to be a token of the species *Elapharus davidianus* in the wild, as there had not been such a species in the wild for roughly 1500 years, if not more. The species that Père David named *Elapharus davidianus* in the mid-nineteenth century was not a wild species, but an immurated species. The individual Père David deer then and now is a token of the species *I(Elapharus davidianus)*. It is, therefore, incorrect to claim that the animal had been saved from 'near extinction', first by Père David himself, and then by the Dukes of Bedford in Woburn Park, if that claim means what it says, namely 'the near extinction of the wild species'. It is simply the case that Père David and the Dukes of Bedford had/have saved the *immurated species, not the wild species, from near extinction.* The deer now in the Dafeng Reserve are, therefore, more akin to feral animals than to the long-extinct wild deer.

It is in accordance with good science to assume that animals with such a long history of immuration are bound to be different in very significant ways from their extinct wild ancestors. In general, we know that captivity brings about changes at all levels – genotypic, phenotypic, metabolic, behavioural. Genetic decline in zoos is commonly observed.[22] I will mention a few instances of other forms of decline/degradation here. As far as birds are concerned, even first-generation captives show deformities in their wings; their inability to fly means that the entire organism is affected, their skeleton, their organs, their plumage, as they fatten on a zoo diet. Captivity undermines the defence instinct; nenes or Hawaiian geese, captive-bred at Slimbridge (UK), were killed by predators when released in the wild. Ungulates in captivity have a lower adrenalin emission, a condition which would disadvantage them if they were to be released to the wild. Captive animals tend to be sexually precocious, and engage in sexual activity more often than their relatives in the wild (probably out of boredom, with nothing better to do).[23]

Conservation scientists tend to overlook or underplay the points made above because they seem to subscribe to what may be called *genetic reductionism* or *genetic determinism*. This is to conceive an animal as no more than its genetic inheritance. Put simplistically, it says that an animal is its genes. As we have already seen, (*ex situ*) conservationists are very careful to ensure that the genes of those captive-bred animals destined for introduction to the wild should reflect the range of genetic variability as found in the wild population(s).[24] This appears to be a fundamental aim and objective by which they judge success or failure in their captive-bred programme. This goal is a necessary precondition for eventual success in reintroducing such animals to the wild but it should, nevertheless, not distract them from appreciating that an animal is more than its genetic inheritance, that it is also the outcome of the complex causal interactions between those genes and their habitat/environment.[25] To put the point more neatly: the animal is not merely its germ-line but its lifeline which emphasises that the animal can only be grasped and understood properly within the interacting contexts of its evolutionary, ecological as well as genetic history.[26] The zoo environment is not the same as the habitat in the wild to which the animal would be released. Hence the captive-bred animal is not the equivalent of its relatives in the wild. If this understanding had been fully absorbed and in the forefront of their consciousness, the scientists involved, for instance, in the Californian condor reintroduction programme would have anticipated the difficulties of reintroduction, instead of being somewhat surprised when they learned about them in an *ad hoc*, trial and error manner. It should also have made conservation scientists wary of claiming that the Père David deer has been saved from 'near extinction in the wild' when that programme has done nothing of the sort.

The lesson about the limitations of genetic reductionism and a full and proper grasp of the zoological conception of an animal, despite experience, remains only partially absorbed as is evidenced by the eagerness of some conservation scientists and commentators to embrace the cloning of animals for the purpose of saving them from impending extinction in the Noah's Ark provided by (*ex situ*) conservation, or indeed, even extending the Ark in recreating long extinct mammals, such as the mammoths *via* the cloning of fossil genetic remains.[27] Recently (August 2004) Indian scientists, including those at India's Central Zoo Authority, have announced plans to save the dwindling population of 300 Asiatic lions (considered as the symbol of India) which live in the Gir Sanctuary in Gujarat and the Indian cheetah (which became extinct in the wild half a century ago) via cloning techniques. For the latter project, the scientists intend to use not a female cheetah, but a female leopard (which unlike the cheetah is not critically endangered in India) to carry the cloned embryo, as Iran appears unwilling to lend India a cheetah for the purpose. Indian scientists are also hoping to get a nucleus containing DNA from the dead skin cells of a cheetah. However, all this excitement on the biotechnological front

appears to overlook the fact that to be able to clone an animal successfully is not to ensure that it can be successfully reintroduced to the wild,
as this presupposes, amongst other things, the availability of a suitable
habitat for the animals.[28] The Wildlife Protection Society of India has
correctly pointed out: 'We are losing forests thanks to highways and
road projects and poachers are killing our tiger population. Cheetahs
need antelope to eat and space to hunt. We do not have enough of
either.'[29]

Education

Hancocks (2001) is one of those zoo professionals who have questioned
the central justification of zoos today in terms of conservation, in particular, *ex situ* conservation via captive breeding. Instead, he advocates
the goal of educating the public as their key justification. He cites
approvingly William Conway's mission statement, uttered more than
40 years ago in 1961:

> Zoo visitors should have the opportunity to learn something about
> each animal's environment through natural habitat displays, to
> explore the mysteries of wild animal behavior, [and] to be informed
> by special displays . . . The justification for removing an animal from
> the wild for exhibition must be judged by the value of that exhibi-
> tion in terms of human education and appreciation . . .
>
> (Hancocks, 2001, p. 110)

However, in spite of Hancocks's unfavourable assessment of zoos in terms
of their central goal of conservation, in particular, of *ex situ* conservation,
it is not obvious that he disagrees in reality too profoundly with those he
criticises, as is evidenced by his endorsement of Conway's position cited
above. Today's zoo professionals and scientists see educating the public as
a key goal, especially when that goal is inextricably entwined with the
central one of conservation. Some zoo advocates go so far as to say that

> zoological institutions only have a right to hold animals in captivity
> if they satisfy one (or both) of two criteria. 1. The animals must be
> part of a managed population with the eventual aim of a 'release-to-
> the-wild' programme in mind. Such a programme may be five years
> or five generations away but it should be our goal . . . All zoos should
> strive to link *ex situ* conservation with *in situ* conservation pro-
> grammes . . . 2. The animals in zoos should form part of a *structured*

conservation-education programme. . . . In the words of Eudley (1995) 'The major function of zoos in protecting biodiversity may prove to be conducting education programs designed to raise the public's ecological awareness.'

(Stevens and McAlister, 2003, p. 99)

Hancocks himself appears to endorse such a stance:

> It is within a wider definition of education that the best and most viable reason for the continuing existence of zoos can be found. They have enormous potential to shape public opinion, to encourage sympathetic attitudes toward wildflife, and to educate the public about ecology, evolution, and wild animals. Zoos can open windows to a world of Nature that people could otherwise experience only via technology . . . It is essential to reach the point where the only zoos allowed by law are those that aim to create respect for wildlife and a desire to save wildlife habitat, by making animal welfare their first priority, by adopting conservation strategies as a central tenet of their operational, budgeting and marketing decisions, and by injecting passion and daring into their interaction with visitors . . . If these changes do not occur, then zoos must surely become increasingly meaningless . . . Zoos have the potential to present holistic philosophies with greater veracity and impact than any other type of natural history institution because they can present and interpret all parts of the story . . . More zoos are becoming habitat based, explaining ecosystems rather than only reciting facts about animals.
>
> (2001, pp. xviii–xix)[30]

If education as understood above is the key, if not the sole justification, of zoos, then we must evaluate it in terms of both its practice and its theory.

The practice, *prima facie*, is immensely successful. Of the 10 000 or so zoos estimated to exist in the world (although, however, they do differ from one another in more ways than one), the World Association of Zoos and Aquaria (WAZA) numbers roughly 2000 as federated zoos, which it considers to form the core of the zoo world, and to which the *WZCS* (1999) is primarily directed.[31] These are also the zoos which are the most likely to have implemented the philosophy of the zoo which Conway, Hancocks and others cited above advocate. Many such zoos (especially those in North America, Europe and Australasia) are rich enough to afford to construct naturalistic enclosures for exhibiting their animals, which is the way recommended by zoo professionals to be the most effective, if not indispensable, means of educating the public about the vital need of protecting biodiversity and the various and different ecological habitats in which wild life exists.[32] As zoos remain hugely popular, their numerous visitors must, therefore, have been exposed to such 'structured conservation-education' programmes. It follows, too, that the visitors should and would have absorbed the information as well as the lessons purveyed by these programmes.[33] But have they, as a matter of fact?

Alas, it appears not to be the case. For a start, visitors in general do not stay for a long enough period of time to absorb all the relevant information; 50 per cent of the visits, it is said, even to Woodland Park (in the United States), last no longer than two hours. According to one set of figures for the Reptile House at the National Zoo in Washington DC, the average time taken to cover the entire house was 9.7 minutes and the average time spent in front of each exhibit was 0.44 minutes.[34] Stephen Kellert, in a US nation-wide survey of 3000 American zoo visitors, found that they failed to identify the basic groups of animals, that they had one of the most limited levels of knowledge and understanding about ecological issues, they scored no higher than non-zoo goers in their knowledge of such matters. Indeed, amongst those with the lowest score are zoo goers – 'Animal-related activity groups with relatively low knowledge scores included zoological park visitors . . .' (Kellert, 1981, p. 14). They appeared to be more concerned with the welfare of the animals they saw in the zoos – that the animals be kept in sanitary and comfortable conditions – than in issues concerning biodiversity, conservation and ecology. 'The zoo-goers group revealed considerable affection and concern for wild animals, but had one of the most limited levels of knowledge and understanding about ecological issues' (Hancocks, 2001, p. 156).[35] The European experience does not seem to be any different. Baratay and Hardouin-Fugier note that

> [s]urveys from Frankfurt to Mulhouse have shown that 80 per cent of visitors questioned claimed to have learned something, but a 1979 investigation noted that they were less sensitive to the need to respect nature than hikers, even after their visits.
>
> (2002, p. 236)

Let us pause at this point to perform a thought experiment. Imagine that the average zoo visitor has all the time available at her/his disposal and is a conscientious seeker of knowledge. What would such an ideal visitor have learnt at the end of a long day at the zoo? It is true that s/he would have read/heard/seen and hopefully understood and digested most, if not all, the material amply available in a variety of forms – whether these be general or more specific information sheets/brochures distributed at the main gate or at the entrance of each enclosure, audio-visual shows in the multi-media centre, special lectures with regard to some aspect of the animals in the zoos, or conducted tours of selected exhibits, and so on. As a result of this kind of educational exposure, the visitor would no doubt be better informed than some other citizen who has not been similarly exposed. However, even under such ideal circumstances, can we simply infer that the combined exposure to the information available and the presence of the animals themselves is either uniquely efficacious, or more efficacious than other methods, for instance, than just simply exposing the person to the information together with the showing of some wildlife films? In so far as evidence appears to be available, such evidence seems to indicate that the zoo experience is neither uniquely efficacious nor more efficacious than other forms of experience and exposure to facts about nature. Zoos are not the only source of information regarding biodiversity and ecology available to the average citizen; it is accessible in numerous other ways. What is unique is the availability of such information presented in a milieu where the animals can be seen in the flesh. Yet there appears to be nothing uniquely efficacious about such a combination of exposure, as the data already cited seem to show.

If education is the key justification of zoos, then zoos appear to be a miserable failure; they have not unquestioningly got their educational message across to the public. Furthermore, let us carry on with our thought experiment and take it a stage further. Imagine that the ideal visitor is not merely a conscientious seeker after zoological knowledge, but also someone who is reflective about such knowledge garnered within the zoo experience. What conclusions would such a thoughtful visitor come to? We have already seen that what is unique about the zoo experience is exposure to the relevant biological/ecological information in the presence of the exotic animals themselves. It is to this unique juxtaposition that the reflective visitor would, therefore, turn her/his thoughts. It would then strike such a person that a severe dissonance – both at the ontological as well as the conceptual level – exists between the message embedded in the explicit texts and that conveyed by the sub-text behind the animal exhibits themselves. The former is about conserving biodiversity and ecosystems/habitats in the wild; the latter is about the presentation of exotic animals as exhibits within a context which is human-designed and human-controlled to promote human ends, albeit including, that of educating humankind in the need for conserving wild species in their wild habitats. The earlier chapters of this book have laboured long over the point that while wild animals in their wild habitats live 'by themselves' as well as 'for themselves', and are the outcomes of the processes of natural evolution and natural selection, captive animals are immurated animals which no longer live 'for themselves', and are biotic artefacts. In other words, what jars is precisely what the *WZCS* considers to be the trump card of zoos, namely exposing the visitor to the 'living animal':

> The most basic level of zoo education is simply the display of living animals. This is the only means by which countless people will ever come into contact with living, wild animals and in a compelling manner become acquainted with elements pertaining to nature conservation.
>
> (*WZCS*, 1999, ch. 4.6)

The animals are indeed living organisms, but they are not wild, in any sense of the word. The thoughtful visitor becomes disoriented on two fronts, the ontological and the conceptual.

Upon further reflection, it might also occur to such a person that zoos themselves have failed to grasp that the dissonance noted exists because they subscribe to a view of animals which may be called *morphological reductionism*. By this is meant that an animal is no more than what it looks – in terms of its size, its anatomy, its shape, its colouring and other visible marks (such as stripes), or whatever other special features are best captured by a painting or a snapshot of the animal. However, zoos claim, of course, that seeing the animal in its living flesh and bones in a zoo is not the same as seeing a painting of the animal or indeed even seeing a stuffed specimen in a natural history science museum. According to zoos, then, what is peculiarly transformative about the zoo experience is to see that animal breathing, moving about or indeed even sleeping, and to be able to straightaway recognise it as a zebra or a lion if one has already seen paintings or snapshots of zebras or lions. However, advocates of the educational merit of zoos appear not to have paused to ask themselves the searching question why this dimension of seeing a live exotic animal should have such peculiar transformative powers.[36] They have just assumed that they exist.[37]

Perhaps the blunt fact of the matter is that it has no such transformative powers, given that the available evidence appears not to support the claim. Its failure as a hoped-for potent educational tool lies precisely in the fact that it rests on nothing more than morphological reductionism, as the sub-text of the exhibits themselves appears to confirm. That sub-text appears to say to the visitor that the exotic animal is 'wild' just because in terms of its outward appearance, it resembles, on the whole, the individual wild animal out there in the wild, be it a zebra or a lion; at the same time, it implies that the habitat of which the animal is an integral part is of no significance.[38] According to this perspective of morphological reductionism, the fact that the captive exotic zebra is kept in a naturalistic, scaled-down, human-constructed caricature of the wild habitat in which the wild zebra lives, that its lifestyle bears no remote resemblance to the life led by the wild zebra in the wild are of no relevance, meaning or importance in identifying it as a 'living wild zebra' as long as it resembles roughly in its outward appearance the individual wild zebra in the wild. In other words, the sub-text confirms that morphological reductionism is what determines that an x is a 'wild x'. This runs totally against the zoological conception of an animal – as purveyed by the explicit texts on biology and *in situ* conservation – according to which the individual animal is an integral part of its habitat, its existence led within its home range as a member of its species is itself the outcome of a long evolutionary history based on the processes of natural evolution and natural selection. It is precisely the simplistic perspective of morphological reductionism embedded in a context of exotic animals as exhibits which stands in the way – theoretically, psychologically/practically – of the zoo experience acting as a transformative educational tool with regard to conservation issues.

A very obvious instance of this dissonance would confront visitors to wildfowl and wetlands reserves such as Martin Mere or Slimbridge (both in the UK). Apart from acting as a valuable point of rest for migratory birds, they also house birds which are not wild but tame, many of them belonging to endangered species.[39] The tameness of mammals under captivity which distorts the reality of mammalian existence in the wild is not as obvious as that of birds. While the uninformed lay visitor may have to glean the fact from the information sheets and other graphics displayed at enclosures to learn about the size of the home range and other details of any particular mammalian lifestyle in the wild, no adult visitor at least, no matter how uninformed, would need to read such brochures to know that it is in the nature of birds to fly, and to fly relatively small or large distances, depending on the species. However, what they see before their very eyes is that the birds do not fly at all, in any real sense of that term. Take the pelican with its enormous wingspans which, if opened in a strong gale, can lift it off the ground, launching it into flight. To render it incapable of flying, at least one of its wings has got to be pinioned, preferably as a chick, to make the operation easier. However, when the bird stretches out its wings to sun itself, for example, it would be obvious to the visitor that one wing has been anatomically tampered with.[40] The taming, in the case of birds, involves in most cases mutilation, which amounts analogously in the case of mammals to breaking one of their legs to make them lame, in order to render them tame and captive.

One could put the point made above in another way. Should the zoo experience have the transformative efficacy attributed to it by its advocates, then the result would be ironical and paradoxical: the transformative efficacy comes from grasping that there is dissonance between the explicit text, on the one hand, about the urgency of the conservation of biodiversity and habitats in the wild and, on the other, the hidden sub-text of simplistic morphological reductionism inherent in the idea of captive exotic animals as exhibits.[41] Furthermore, grasping that dissonance would logically lead to the realisation that the justification of zoos in terms of their educational goal has no basis either in practice or in principle.[42]

Once attuned to the realisation that dissonance is inherent in the zoo experience, the thoughtful visitor would begin to find evidence of it in more ways than one. It suffices just to list a few examples, which are ultimately also derived from the central one already elaborated upon. Zoos are known to pair or group animals in what may be called an anomalous manner. For instance, most species of small felids are solitary in the wild; zoos, however, routinely exhibit them in pairs or trios for a variety of reasons, one of which is to provide the cat with company, an arrangement endorsed as part of the zoos' programme of environmental enrichment.[43] Instead of pairing animals or forming groups out of animals belonging to the same species (conspecific pairing), zoos sometimes form pairs or groups from different species (xenospecific pairing).[44] Zoos again justify this anomaly in the name of enrichment. The complex elephant social systems in the wild are never replicated in zoos, at least in the European zoos studied and cited below. Instead of the extensive family-based structure of wild herds, most of the elephants in zoos are held in small groups of unrelated elephants, consisting mostly of adults with a few infants or juveniles. In the wild:

[t]he matriarch leads the herd, her dependent offspring, and adult daughters with their immature offspring including pre-pubescent

males – family units on average have between six and eight individuals (range 2–40) amongst Asian elephants and four to 12 (range 2–29) amongst African savannahs. These family units associate in 'kin' groups, consisting of up to five families and sometimes in turn form part of a clan consisting of several hundreds of individuals. Males either roam around alone or associate together in bull groups when not sexually active. In zoos, males and females are housed apart; in extensive systems in Asia, elephants are kept in mixed groups of between 51 to almost 3000, which approximate more to conditions in the wild. Female elephants in European zoos are therefore kept in far smaller groups compared to the wild and extensive systems. A solitary state is far more common in wild adult males, but they are also often observed in small groups when not sexually active, a situation not replicated in zoos.

(Clubb and Mason, 2002, ch. 5)

In the wild, the ratio of male to female is 1:1; in zoos there are more females than males, the ratio is 1:4.5 for Asian elephants and 1:3.4 for African elephants. In the wild populations, infants (under 5 years of age) were by far the most common age class; in (European) zoos, the most common age class does not comprise infants, but adults aged 30 to 34 years in the Asian populations and 15–19 in African populations.[45]

These anomalies distort the reality out there in the wild.[46] The mismatch, when noticed by the conscientious and reflective zoo visitor, would only serve to disenchant the observer rather than enhance the zoo experience from the point of view of its expected transformative powers as enunciated by zoo advocates whose primary mission is education.

It is worth labouring the crucial point yet again that zoos, far from showing respect for animals-in-the-wild and their behaviour in their wild habitats, necessarily present their captive exotic animals as mere exhibits, whose behaviour has nothing to do with behaviour in the wild. The behaviour of immurated animals as exhibits implies that behaviour in the wild is considered as an irrelevance to a proper understanding of what constitutes the wildness of animals, or that it is meant to be taken seriously as a way of understanding animal wildness; in reality animal behaviour in zoos is, at best, nothing more than an obscene caricature of behaviour in the wild. Hancocks has, indeed, put his finger on the point when he writes:

Visitors often do not see zoo animals as wild animals because they are not presented as wild animals. They play with beer barrels, bowling balls, and plastic milk bottles. They feed from piles of chopped carrots, heaps of premade chows, beef patties in stainless steel dishes.

(Hancocks, 2001, p. 249)[47]

Zoos proudly present such innovations in animal activity (playing with plastic toys) as laudable and successful attempts to enrich their lives; as captive residents, they would otherwise die of boredom and engage in stereotypic behaviour, upsetting their visitors into the bargain.

Shepherdson, a leading expert and exponent of enrichment programmes for captive animals writes:

Chimps at zoos are furnished with artificial termite mounds. Instead of actual termites, the chimps plunge their sticks into gobs of treats such as mustard, pie fillings, ketchup, apple sauce and other goodies they love . . . They are doing what they do in the wild, only the taste results are different . . . Keepers try to stimulate animals' intellects, making them want to constantly explore their cages for potential food and fun with: Smells, such as spices and animal scents, are placed on logs or rocks in exhibits. Large, tough polyethylene balls go into the cages of our tigers. They like to bang them around or pounce on them as if they were prey. Our tigers . . . hunt for fish and dry meat that keepers hide in logs, giving them a chance to smell and forage for their food as they would in the wild. Bears also get the 'log' treatment, hunting in them for bone, raisins and nuts . . . Occasionally the polar bears receive lumps of ice containing frozen food items. Polar bears are also given plastic balls and tubs to manipulate, play with and ultimately destroy.

(Shepherdson, 2005)

Hancocks (cited above earlier) is quite right. Imagine what children taken on an educational trip to a zoo would take away as lasting impressions of what such 'wild' animals do in their spare time! Children exposed to such scenes would be highly unlikely to make the intellectual as well as psychological leaps from the evidence before their very eyes to the story revealed in the literature about truly wild animals and their behaviour in the wild and thereby benefit from the postulated transformative powers inherent in the zoo experience. The transformative powers could not be realised as a result of the dissonance between what they see and what they are told to see/read.[48]

The postulated transformative powers can never be realised, no matter how much effort zoos may make to reduce the dissonance noted. In the 1970s, Woodland Park Zoo (United States) began feeding whole carcasses of sheep and goats to the big cats and of rabbits and chickens to the small cats. Outraged citizens protested and the zoo stopped the innovation at once.[49] Danish zoos have spent a lot of time and energy in preparing their visitors over the years to accept more naturalistic diets; for instance, they have been able to feed horse or cow carcasses to their lions without the Danish citizenry being outraged. Whenever the zoos get a fresh carcass, the visitors are treated to a talk by a keeper and an educational officer to explain what is happening. When presented with a whole large carcass, the lions would behave more like lions in the wild – for instance, the male lion is the first to go up to it to feed, then followed by the cubs and the female. Furthermore, unlike the normal zoo food, such a carcass takes more time to consume which makes it approximate more to consuming prey in the wild.[50] However, there is a limit to such 'enlightened acceptance'. No federated zoos at least in the Western (economically) developed world would dare to allow a large predator, such as a lion, or a tiger hunt a live prey such as a deer, even if this were practical, although some zoos may feed their large carnivores with small prey, such as a live chicken, behind the curtains, so to speak, in non-exhibit enclosures.[51] This limitation is but a reflection of the essential fact that a zoo is a human cultural space, where human values prevail and in which captive animals must be treated in ways consonant with these values. As the thrust of this book has argued, zoo values are just not compatible or consonant with the reality of life as led by wild animals-in-the-wild. There is, therefore, bound to be dissonance between the two worlds, that of the zoo and that of the wild, and into the gap falls irretrievably the so-called potential transformative powers of the zoo experience.

Conclusion

This chapter has argued that none of the three justifications of zoos deemed to be serious can stand up to critical scrutiny. The flaw lies crucially in the inevitable dissonance – fundamentally at the ontological level – between, on the one hand, wild animals, their behaviour in their wild habitats within their own home range and, on the other, immurated animals and their behaviour as exhibits in zoos which are human-designed and human-controlled institutions serving human ends, whether these be furthering scientific knowledge, conservation or education.

In the domain of scientific research, it would be bad science, into the bargain, to extrapolate uncritically from data achieved with immurated animals to very dissimilar contexts elsewhere. In the domain of conservation, we have seen that captive breeding as part of *ex situ* conservation is incompatible – ontologically and practically – with the conception of zoos as collections of exhibits open to the public; as such, these activities are best hived off to other institutions which may be called captive breeding conservation centres. The definition of 'zoo' cannot be stretched indefinitely to cover two fundamentally diverse forms of activities, without creating confusion in thought and in policy. In the domain of education, there is little or no evidence to back the claim that zoos are uniquely efficacious in driving home ultimately to the public the need for the conservation of wild animals in their wild habitats. Furthermore, the unique potential transformative power of the zoo experience, as claimed by those zoo advocates who see zoos as the crucial tool in their mission to educate the public in matters of biodiversity and ecology, cannot and can never be realised, given the distinct mismatch between what zoo visitors see with their very own eyes, which is at best a caricature of wild animals and their behaviour, and at worst is a denial of the relevance and significance of wild animals and their behaviour in the wild. Immurated exotic animals and their behaviour exhibited in zoos are the ontological foil to wild animals-in-the-wild and their behaviour in their natural habitats within their normal home range.

10
Justification Deemed Frivolous

The last chapter has critically challenged three justifications of zoos deemed serious and found each wanting. We turn now to the fourth justification, usually considered today to be less respectable, and therefore sometimes not even deemed worthy of mention by serious-minded zoo professionals and advocates. It is recreation.

Why frivolous?

It is not easy to comprehend why the recreational justification of zoos is held in somewhat slight regard in certain quarters, when it is obvious that zoos play a vital social role in providing recreation and entertainment, and that zoo-going is one form of leisure activity competing against others for the attention and attendance of the public, such as watching television, playing computer games, going to sporting/musical/theatrical events.[1] According to the statistics, zoos appear to be doing exceptionally well in attracting punters, so to speak, to their gates. We have seen in the last chapter that there are more than 10 000 zoos world-wide. However, as figures of attendance for all of them are hard to come by, it would be best to confine the numbers only to the 1200 or so federated and accredited zoos which are said to constitute the core of the zoo-world today; nevertheless, they still come to the staggering total of at least 600 million, about a tenth of the entire population in the world per year.[2]

Zoos are big draws and big business. Why? *Ex hypothesi*, it is not because zoos are institutions which carry out scientific research or captive breeding conservation programmes, for the simple reason that these two forms of activities are not open to public visitors. So that leaves education; however, in the last chapter, we have cited statistics to show that the average time spent before enclosures and the attention-span on the part of visitors do not, as a matter of fact, permit them to pause to read, absorb and digest the detailed material made available to them about matters of biodiversity, ecology and conservation. It then seems reasonable to infer that zoos are big draws precisely because the public, flocking to visit them, do so for recreational/entertainment reasons, and not primarily for the so-called noble motive of educating themselves about the wild and the urgent need to save wild species and their habitats from extinction.

Zoo-going is, undeniably, for a substantial part of the world's population, a congenial and agreeable form of family outing, rather like a picnic in some park or beauty spot in the country. It is, indeed, superior to these other examples of outdoor recreation, because when the weather becomes inclement, one can take refuge in the indoor enclosures. It is true museums provide shelter, too, when it rains or hails. But museums, on the whole, do not welcome children, and even when they do, the children, like those in Victorian times, are expected only to be seen but not heard. Museums are hushed places; for parents to keep their children under severe control so as not to disturb the quiet reverential ambiance, is so stressful that museums are best avoided. In a zoo, children are welcome; they can run around, and fellow visitors appear on the whole not to mind when children behave as children tend to do, as zoos, in general, unlike museums, are not places which expect visitors to keep a respectful and reverential silence in front of the objects they have come to see. Furthermore, children do get excited when they see a charismatic animal such as a tiger or a chimpanzee, whereas it would be a very odd and unusual child indeed to get excited in an analogous way and to the same extent when s/he is taken to stand in front of the *Mona Lisa*. So parents, being sensible, opt to take their families to zoos instead.

Shared presuppositions

Zoo-visiting as a form of recreational activity demonstrates that zoo visitors act upon the understanding that zoos are essentially collections of exhibits consisting of live exotic animals, open to the public. For them, the crucial aspect of the exhibits is that the animals are alive, not dead and stuffed, as are the exhibits they would find should they visit a natural history of science museum. In this, they are at one with zoo professionals, who also claim that the unique attraction of zoos over these other establishments exhibiting animals is that their exhibits are indeed live animals.

As long as the animals ostensibly breathe, move about, flick their tails or their trunks, make other movements such as climbing poles or frames, or make vocalisations such as growling or screeching, these appear to be sufficient to please the punters. At the same time, the animals must look like those they have seen pictures or other images of, so that they can immediately recognise the exhibit as a lion or an elephant, and so on. In other words, they subscribe implicitly, as zoos do, to the view referred to, in the last chapter, as morphological reductionism – basically, what an animal is, is what it looks like.

Zoo visitors, on the whole, appear indifferent to the ontological status of the exotic live animals before their very eyes; in other words, they are not particularly worried or concerned whether such animals are wild as zoos claim they are, or whether they are truly wild. They appear to side with the zoo interpretation of 'wildness', namely an animal is 'wild' if it looks like the wild animal in the wild existing in its own home range. They are complicit with zoos in accepting for their purpose that the animal exhibits are 'wild'. Those members of the public with ontological scruples of the kind advanced by this book are few and far between; they either keep away from zoos altogether or if they do visit, they do so not on the understanding that they would encounter truly wild animals. As such, they are not likely to ask for their money back on the grounds that zoos have violated the Trade Descriptions Act; nor, indeed, would they necessarily succeed were they to sue, because of their prior understanding that zoos do not in reality exhibit wild animals or animals-in-the-wild.[3] Zoos do not need to lose sleep over such potential litigation; the majority of their visitors are happy and satisfied customers.

As already remarked upon in the last chapter, their visitors would indeed be unhappy or dissatisfied should the animals appear to them to be mangy, undernourished, obviously unwell or living in unsanitary conditions. So long as the animals look well-fed, not ill or diseased, the visitors would have nothing to complain about. In this respect, the (environmental) enrichment programme that today's zoos are committed to fits in absolutely well with the expectations of zoo visitors. As one such expert has correctly remarked, animals in captivity easily succumb to boredom, as they have nothing really any better to do to keep themselves occupied than to engage in stereotypic activities, such as nibbling their tails, shaking their heads endlessly, pacing for no obvious good reason, or throwing their faeces about.[4] These would distress the visitors; zoos rightly attempt to eliminate them via their enrichment programmes, for the sake of their animals and of their customers as well as of their own financial interests.

We have already observed that the majority of visitors are not particularly concerned either about the ontological status of the animals they see or the explicit mission of zoos to educate them in matters of zoology, ecology or biodiversity, and as their aim is to get some entertainment looking at how live animals behave in zoos, it would not matter to them if the animals behave in ways which bear no resemblance to what real wild animals actually do in the wild. They are not perturbed that zoo animals cannot hunt or forage, or roam large distances. On the contrary, they can be mightily amused by the sight of the polar bear trying to submerge itself in the tiny pond of ice-cold water provided in its enclosure, or to catch with its paw the fish in it, or to find a chocolate encased in a lump of ice, or to play with plastic balls which it then crushes with its paws. They can be agog with delight to see a chimpanzee using a stick to poke into the artificial termite mound, not to fish out termites as their relatives in the wild would as their proper food, but supermarket treats such as mustard, pie fillings, ketchup, apple sauce and other processed foods that they themselves and their children might well be munching at that very moment as they stand in front of the chimpanzee trying to get hold of similar goodies. They are not remotely concerned with the fact that in the wild, the chimpanzee and the polar bear would not be eating the kind of food or acting in the kind of way they are constrained to in their enclosures. Indeed, the greater would their delight be, should the experts on enrichment come out with new ways of making the lives of the animals less boring and more entrancing for them. For instance, chimpanzees can be trained to use computers; they appear to enjoy themselves tremendously in watching computer games.[5] It is obvious to the visitors that chimpanzees in the wild would show no interest in such gadgets or in games and would be preoccupied with a very different set of activities altogether; however, this realisation would not stand in the way of their enjoyment in seeing a chimpanzee in a zoo doing precisely that.

Visitors to some zoos would find it just as amazing and amusing to buy a special ticket to be allowed into a special enclosure to take tea with some chimpanzees or have a so-called 'wild breakfast' in the company of orang-utans.[6] They know, of course, that chimpanzees in the wild would not be sitting down with them at a table while they drink tea or have breakfast; but the zoo is not wilderness and the chimpanzee or orang-utan hosts are not truly wild. So it is fun. The chimpanzees and the orang-utans appear to be enjoying themselves; it is far better that they sit at a table with humans than that they tear out their own hair in frustration. Some zoos also offer facilities for nuptial and other receptions, using their animals as backdrop to enhance the pleasure of such occasions. One could choose one's favourite animal amongst those on offer.[7]

The logic of recreation may be seen in its most extreme as well as its purest form in the kind of entertainment which has been honed to perfection and which pulls in the crowds at Sea World, where Shamu, the whale, does amazing turns under the command and control of its trainer.[8] Those who want to see what whales in the wild actually do are not the people to turn up at Sea World; however, the spectators of the show know full well that whales in the ocean do not behave in that fashion, obeying human commands. What thrills them is precisely the way the whale reacts to such commands; what they marvel at is also the skill with which the trainer has got the whale at his/her beck and call.

Zoo visitors equally enjoy the novelty, charm and thrill of seeing zoo keepers exercising/walking a cheetah or a tiger, an experience unique to zoos. Naturally, such a sight would never be found in the wild; nor would one find it in one's suburb. The cheetah or tiger is definitely not a dog, yet its keeper is taking it for a walk in the same way as they themselves, in their role of suburban dog owner, walk their own dog. It is tame, yet it looks wild. Herein lies the peculiar fascination which comes from the conjuncture of frisson and amusement.

Conclusion

In other words, zoo visitors are tacitly aware that a zoo is a human cultural space and that the animals exhibited within it are expected to conform to human practices, norms and values. They know, then (even if they do not articulate the thought in quite this manner), that the exhibits are immurated animals and are biotic artefacts, responding to and interacting with the stimuli orchestrated by their human controllers and managers. They, therefore, implicitly know that the animal in front of them is not a token of the wild species. The individual zoo orang-utan is not a member of the species (*Pongo pygmaeus*), but of the species *I(Pongo pygmaeus)*.

This is to say that they know what the score really is and with that tacit awareness, they can sit back, relax and enjoy their time at the zoo in the presence of the exhibits as a form of recreation and entertainment. For those who care about the welfare of animals, they may even prefer to visit a zoo rather than a circus, as zoos, on the whole, are more successful than circuses in convincing them that the animals under their care do not suffer, but are happy and comfortable in their enclosures.

Their satisfaction and fascination, as we have already observed *en passant*, comes from the tension they perceive between the fact, on the one hand, that the exhibits look wild, as they look just like the animal-in-the-wild (their tacit commitment to morphological reductionism), and on the other, their tacit appreciation of the fact that they are not wild. This adds to their thrill of being in the presence of something which could be potentially dangerous and threatening, such as a truly wild tiger or lion while knowing that the exhibits are not truly wild, that the animals would not be able to cause harm to the visitors, since they know that the zoo is a human space made safe, but interesting and entertaining for humans.

It may well turn out that ordinary zoo visitors are more insightful about the true nature or 'essence' of zoos than the noble-minded advocates of such institutions in the name of education-cum-conservation. It is no wonder that such missionary zeal, on the whole, leaves them unaffected and passes them by – rightly, they concentrate on what they have come to zoos for, namely, to see and be amused/entertained by live exotic animal exhibits in pleasant naturalistic simulated enclosures. Good wholesome recreation and fun to be had by one and all in the family. This is not to be sneered at, especially when the satisfied customers appear not to have been taken in by the zoos' own mistaken and misleading spiel that the animals they exhibit are wild.

11
Philosophy and Policy

In Chapter 1, we identified five conceptions of what counts as an animal, namely the lay conception, the zoo conception, those presupposed respectively by Singer and Regan in their philosophy of animal welfare/animal rights, and the zoological conception. The subsequent chapters investigated the fundamental issue concerning the different ontological status of zoo (captive) animals on the one hand and that of animals-in-the-wild on the other. It is now time to look more closely at this fundamental issue in the context of the five conceptions raised in Chapter 1. There are some surprising conclusions which emerge from this examination.

The zoological conception and its ontological presuppositions

This conception of what counts as an animal and what is a wild animal need not delay us for long; it is this conception which has inspired precisely the exploration of the difference in the respective ontological status of exotic zoo animals as exhibits and animals-in-the-wild. The ontological perspective of this book, therefore, coincides with those implicit in the zoological conception concerning animals-in-the-wild. A quick summary should suffice:

- Animals include all animals – as identified and classified by the current consensus of the scientific community – whether vertebrates or invertebrates, of which there are over a million known species at the least.
- Animals-in-the-wild can only be properly grasped and understood when they are seen as the outcome of the processes of natural evolution and the mechanism of natural selection within their natural

118

home ranges, interacting in a causally complex manner with their habitats and ecosystems of which they are an integral part. Their morphology, their genetic inheritance, their behaviour are the outcome of such processes of reciprocal causation.

- Animals-in-the-wild live 'by themselves' and 'for themselves', pursuing their own trajectories, independently, in principle, of human design, manipulation and control.

The zoo conception and its ontological presuppositions

These constitute the *ontological foil* to the zoological conception of animals-in-the-wild:

- Zoos claim to keep animals, but in reality, they house only a tiny portion of all the species of animals known to science; these are usually confined to the vertebrates, especially large- and median-size mammals. This extreme selectivity in their reference implicitly denies the status of being animals to those hundreds of thousands of species excluded.
- Zoo animals are necessarily exotic in both the technical as well as lay senses of the term – they are animals which have been removed from their natural home ranges, transported to other climes in other parts of the world, and they are, on the whole, considered to be charismatic, rare, especially cuddly or appealing to humans because of their looks or behaviour.
- Zoos are defined as collections of animal exhibits open to the public; this definition encapsulates in a nutshell the difference in ontological status between zoo animals and animals-in-the-wild.
- As exotic exhibits, the animals necessarily lead lives in environments which are totally different from those led by animals-in-the-wild, because a zoo is a human cultural space within which the animals under captivity are expected to conform to the purpose and norms of humans who control and manage them.
- As immurated animals, they are no longer subject to the processes of natural evolution; they are, therefore, biotic artefacts. As such they are the *ontological foil* to the wild animals living in the wild.
- As immurated beings, they do not live 'by themselves' nor do they live 'for themselves' since they live to fulfil the destiny that humans have ordained for them.
- Zoos maintain that the animals under their care and control are 'wild'; yet that claim cannot survive critical scrutiny. It is, therefore, both a conceptual and an ontological mistake for zoos to hold that immurated

animals are wild animals; they are not wild and are not tokens of any wild species; they are tokens of what may be called immurated species.

The lay conception and its ontological presuppositions

According to the deconstruction pursued in the last chapter, the zoo-going public appears to have an intuitive grasp of the ontological issues embedded in the zoo conception of animals as exhibits. Surprisingly, one is able to infer from its attitude certain (implicit) tenets and construct a more or less coherent and, indeed, even a sophisticated conception of zoos and their justification.

* The majority of zoo visitors regard zoos as places of recreation rather than serious education.
* The zoo-going public agrees wholeheartedly with zoo management/philosophy that zoos are essentially collections of animal exhibits.
* The animal exhibits constitute the focal point and source of their entertainment. As such, visitors expect zoos – and zoos happily oblige – to exhibit, by and large, charismatic, rare animals which, on the whole, tend to be vertebrates, especially large- or median-size mammals. These exhibits are exotic in the lay sense of the term.
* The zoo-going public intuitively grasps that, although the animals may look like animals-in-the-wild (morphological reductionism), they are exotic (in the technical sense of the term) and are not really wild, as they live in environments, geographically far from their home ranges, which are totally different from habitats in the wild.
* Visitors do not, therefore, expect animals in captivity to behave in the same way as animals-in-the-wild.
* The zoo-going public, nevertheless, finds the behaviour of animal exhibits within their human-designed and human-controlled space fascinating, amusing and pleasing, provided it is convinced that the animals themselves do not suffer pain in being encouraged to display such modes of behaviour while passing their time in confined space. It implicitly and happily accepts that the programme of enrichment is a good thing, as enrichment makes the animals less bored and, therefore, also less boring to watch.
* *Au fond*, the zoo-going public appears to agree with the implicit ontological stance behind the zoological conception of animal which this book makes explicit, namely that zoo animals are immurated animals, and therefore are biotic artefacts, and that they are not tokens of any species in the wild.

Philosophy of animal welfare

This philosophical perspective is neutral as to whether the animals are truly wild or tame, domesticated or undomesticated, as its remit is confined solely to sentience and the prevention of suffering whenever possible. As such, it is agnostic with regard to the ontological status of zoo animals. The morality of sentience demands that all sentient animals – in particular, mammals, in which zoos take an inordinate interest – ought to be treated in ways which do not cause them pain (or cause them less pain), especially physical pain. Such a philosophy chimes in admirably with the philosophy of enrichment which today's progressive zoos practice, as such a programme does have at its centre the welfare of the animal. That is why while animal welfare activists picket animal laboratories (which breed animals for use in scientific experiments which often, if not invariably, involve pain), they do not stand outside the gates of well-run zoos, which are perceived to do all that can be done to reduce, if not eliminate, suffering in animals under their care and control. However, organisations such as the Royal Society for the Prevention of Cruelty to Animals (RSPCA) do protest and demonstrate against poorly-run zoos which ignore the health and well-being of the animals in their charge.

Philosophy of animal rights

Unlike the philosophy of animal welfare, the relationship between this philosophical stance and that pursued in this book regarding the ontological status between zoo animals and animals-in-the-wild, is somewhat more complicated. One needs to distinguish between two contexts. The first involves the commonly perceived situation regarding zoos, that they house wild animals, so to speak. It is safe to assume that those today who subscribe to the morality of animal rights, also share this perception; it, therefore, follows that their philosophical view about animal rights is distinctly incompatible with the philosophy of zoos and zoo management including that of enrichment, as zoo practices and procedures make life difficult for animals which are 'subjects of a life'. The morality of animal rights sees the denial of the freedom to roam, to lead lives as animals do in the wild as amounting to slavery. On this view, slaves are slaves, whether they are slaves suffering and miserable in chains, or contented slaves living in safe, sheltered and luxurious accommodation – the equivalent of the Ritz Hotel – and are cared for and looked after by the best medical consultants in the world and their team of health workers. The Born Free Foundation is not the RSPCA; when pressed, it would surely argue that freedom to lead independent lives which bring inevitably in their wake, hunger, cold, thirst, disease and early death is better than living in security from famine, disease and early death but at the price of being unfree. A gilded cage is still a cage. Zoos, in principle, are morally dubious institutions and, therefore, unacceptable.[1] However, the second context is a potential one which involves the rejection of the commonly held perception that zoo animals are wild animals in the light of the ontological arguments set out in this book – those who uphold the morality of animal rights may come to agree that zoo animals, being immurated animals, are a special type of domesticant, especially those animals which are the offspring of immurated ancestors, and therefore are not proper subjects for liberation from zoos, any more than cats and dogs are proper subjects for emancipation. Those who react in this way may then be said to belong to the moderate wing of the animal rights movement; those who consider domesticants of all varieties, whether classical or the newly identified immurated type, to be the proper target for liberation, would not, naturally, be impressed by the ontological standpoint advocated by this book.

Policy conclusions

From the ontological vantage point of this book, it transpires that zoos, in principle, are not necessarily unacceptable institutions; indeed, they are a positive cultural asset provided certain important caveats are entered:

• First of all, one needs to draw attention to the fact that the ontological standpoint of this book is in a sense, on the surface, compatible with the philosophy of animal rights with regard to zoo animals. It, too, deplores the fact that animals directly captured from the wild are denied freedoms; however, it regards animals born and bred under captivity in a totally different light, as their degree of artefacticity is much greater than that of the former category of animals, which have not been transformed so thoroughly into biotic artefacts.[2] In other words, the ontological perspective belongs to a totally different type of discourse from that of the philosophy of animal rights – the central value of the latter rests on a particular normative ethical theory, that of rights, while the central value of the former is derived from the premise that animals-in-the-wild have come into existence, continue to exist and go out of existence, in principle, independent of human design and control. These are different in ontological status, and therefore, possess a value different from animals whose existence is shaped and controlled by human intentions and goals. The wildness of animals-in-the wild is an intrinsic value which has nothing to do with the goals and purposes of humans; in contrast, the value of immurated zoo animals, which are biotic artefacts, is primarily instrumental, to serve human intentions and purposes. As a result, unlike the morality of animal rights which condemns all zoos in principle, the ontological stance is not *per se* inhospitable to the idea of well-run zoos from the point of view of animal well-being. In this aspect, it is compatible with the philosophy of animal welfare; zoos which are human cultural spaces are, therefore, quite morally acceptable if they are run on the 'humanitarian' lines endorsed by a philosophy based on sentience. It is precisely the emphasis of this study on the ontological difference in status between animals-in-the-wild and animals in zoos which enables it, as a result, to adopt a much more nuanced attitude to zoos.

- It also follows from the above that zoos must not, however, claim that animals under captivity, living at best in simulated naturalistic environments, are wild, and recognise that they are not tokens of species in the wild.
- Once these crucial concessions are made, there is no ontological objection to zoos, provided that zoos are prepared to acknowledge, in consequence, that immurated animals are in fact domesticated animals. One objects neither to immurated animals nor to traditional domesticants; one could only legitimately complain should their ontological status be misrepresented by the claim that the former are 'wild animals in captivity' when, in reality, they are biotic artefacts.
- Hence, the ontological stance endorsed by this book is sympathetic to the attitude of zoo visitors who, by and large, are very comfortable with the reality they are confronted with, namely, that they are looking at, and being entertained by a unique kind of animal, that is to say, immurated animals which are found only in zoos, and that such animals do behave very differently from animals-in-the-wild. For them, this uniqueness is what precisely constitutes the zoo experience.
- Zoos are mistaken in believing that their true goal is not recreation, but some other more high-minded ones such as education about animals-in-the-wild for the purpose of their conservation, as well as *ex situ* conservation which zoos undertake through introducing certain of their captive-bred animals to the wild. Ontologically speaking, zoos are simply not the right space for the pursuit of such aims, noble though the goal of educating the public about biodiversity and ecosystems in the wild undoubtedly is.[3] *The World Zoo Conservation Strategy* (1993) and the European *Union Zoos Directive* (1999) are essentially misguided in the goals and the priority they have set out for zoos.[4]

- Zoos, in the long run, appear to be ideally positioned to create new immurated species in two ways: (a) out of extant wild species through the sheer fact of sustaining generations of captive-bred animals; and (b) more radically through modern biotechnological techniques, of either re-creating extinct species such as woolly mammoths via cloning, relying on fossil DNA, or entirely new chimeric animals/ species. Chimeric technology has already permitted the creation of the geep or shoat (a chimera with a sheep and a goat as genetic parents).[5] Such novelties are bound to thrill zoo visitors of the future and would be sure-fire draws, should zoos choose to embrace this route.

  The usual arguments which are justly mounted against using such techniques as part of *ex situ* conservation would no longer apply, as the products of these procedures are intended to be immurated animals which would be tokens of new immurated species. For instance, one argument against cloning extinct animals in the context of conservation is that such clones would have no suitable habitat to be released into the wild, and even if such a habitat could be found, they would not necessarily know how to survive, as we have earlier seen in the analogous case of the Californian condors. However, as immurated animals, they would never leave the zoo environment and never be expected to fend for themselves. Nor would they be expected to found (genetically) sustainable populations in the wild.

  To clone an extinct animal would require using an extant female animal belonging to a different species to provide the egg as well as to act as surrogate womb – this means that the clone, in reality, is a hybrid, as the mitrochondrial DNA remains in the egg in spite of having its nucleus removed. The contribution of mtDNA to the development of the embryo and later the animal could not be ruled out; nor could the contribution of the surrogate womb be ruled out on scientific grounds. However, in the future zoo context of creating artefactual biodiversity, none of these objections would be valid.[6]

- Zoos, once emancipated from the myth that their collections of exhibits are 'wild animals' may also choose to embrace the strategy of artificial selection of certain features of animals in their keeping, deemed to be attractive to zoo visitors. This would, indeed, enhance their recreational value. For instance, zoo visitors find albino tigers fascinating. At the moment, reproduction of such animals in zoos is frowned upon because they are said to be 'freaks' in nature, that is to say, they occur very rarely; therefore, it would be wrong for zoos to reproduce them because this would mislead zoo visitors about the true state of affairs which obtain in the wild.

Conclusion

Before setting out a few succinct conclusions to be drawn from the philosophical/ontological exploration of zoos, it may be useful to remind the reader of the details embedded in the distinction between animals-in-the-wild on the one hand and immurated/zoo animals on the other.

'Wild' versus 'tame/immurated'

Wild	Immurated
Not habituated to human presence; flight tendency and distance deeply engrained	Tame: Habituated to human presence; flight tendency overcome and flight distance overcome/ reduced through training procedures
Roaming within ancestral/natural home range	Exotic: removed from home range, transported to different geographical/ climatic locations
Existence led in natural habitats	Existence lived in miniaturised, simulated naturalistic environments in exhibit enclosures during zoo opening hours but otherwise often in cages in non-exhibit areas for the rest of the day/night
Daily life dictated by need to get food/shelter, to run away from danger	Hotelification: 'bed and full board' with 'room services'
Diet: hunting other animals in the case of carnivores; foraging for plants in the case of herbivores	Diet: scientifically/nutritionally prepared food pellets, chopped up meat from the carcasses of domesticants such as horse, cattle, supplemented by some foraging in the case of herbivores

(Continued)

Wild	Immurated
Seasonal activity: males actively seeking and fighting for suitable females to mate	Exaggerated sexual and mating activity; mating partners determined by zoo managing policies which include techniques such as *in vitro* fertilisation
Vulnerable to hunger/thirst, wounds/injuries, disease leading to early death	Totally protected from hunger/thirst, wounds/injuries, disease leading to early death
Vulnerable as prey to dangers including wounds/injuries and often death inflicted by predators	Neither prey nor predator and the 'ills' thereof, including early death, exist
Self-medication in certain cases is known to science	Full panoply of veterinary/medical care and services to prevent disease and early death
Life of (usually) females taken up largely with looking after their cubs, teaching and initiating them to become independent mature adults	Nurturing tasks taken over in the main by zoo carers and keepers; teaching tasks are redundant given hotelification and medication
Subject to the processes of natural evolution and the mechanism of natural selection	Suspension of natural selection: human design, control and management act as substitute, involving a form of artificial selection, leading to domestication (though not in the classical sense of the term)
Naturally occurring beings; beings which live 'by themselves' and 'for themselves'	Biotic artefacts which exist neither 'by themselves' nor 'for themselves'
Pursue their own trajectory, fending for themselves	Existence ordained by zoo managers; looked after in all ways by zoo carers and keepers
Exemplify the thesis of intrinsic/immanent teleology	Exemplify the thesis of extrinsic/imposed teleology
Individual animals are tokens of a wild species	Individual animals are not tokens of any wild species, but of novel immurated species
Individual animals form part of a natural population within a certain habitat	Individual animals are exhibits and form part of a collection of animal exhibits open to the public
'Wild animal in their natural habitat': antonym and ontological foil of 'tame'/'domesticated'/'immurated' animal	'Tame', 'immurated': antonym and ontological foil of 'wild animal in their natural habitat'

Summary of ontological implications for policy making

• Zoos do not house naturally occurring wild animals; nor does it make sense to say that they house 'wild animals in captivity'. Instead, zoos house tame, immurated animals. The latter, as biotic artefacts, are the *ontological foil* to wild animals in their natural habitats.

• Zoos should not, therefore, be expected to take on tasks for which they are ill suited, such as *ex situ* conservation as well as education for conservation.

• Zoos through their collections of animal exhibits are uniquely placed, as a human cultural space, to provide recreation and entertainment to their visitors.

• Zoos, far from protecting threatened wild species from extinction as a modern-day scientific Noah's Ark, are well placed to be the creator of new species of animals in the long run; however, these are immurated species. They add to biodiversity; but the biodiversity is not that of extant naturally occurring, wild species but novel artefactual species. (For details of the differences between natural and artefactaul biodiversity, see Lee (2004).)

• In the light of the above, this book throws out a challenge to the World Association of Zoos and Aquariums (WAZA) and the EU to reconsider their justifications for zoos in terms of *ex situ* conservation and education-for-conservation.

Appendix: Environmental Enrichment or Enrichment

This short appendix is not meant to be a comprehensive account of the zoo management technique called either environmental enrichment or, more simply, enrichment. Instead, it aims only at assessing it from the ontological standpoint adopted in this book, namely, that zoo animals are immurated animals, a new type of domesticants in the making.

Let us begin by stating what environmental enrichment is through the words of one of the world's leading experts on the subject:

'Environmental enrichment is a process for improving or enhancing zoo animal environments and care within the context of their inhabitants' behavioural biology and natural history. It is a dynamic process in which changes to structures and husbandry practices are made with the goal of increasing the behavioural choices available to animals and drawing out their species-appropriate behaviours and abilities, thus enhancing their welfare. As the term implies enrichment typically involves the identification and subsequent addition to the zoo environment of a specific stimulus or characteristic that the occupant/s needs but which was not previously present. (American Zoo & Aquarium Association, 1999)'. In practice, this definition covers a multitude of innovative, imaginative and ingenious techniques, devices and practices aimed at providing adequate social interaction, keeping animals occupied showing an increased range and diversity of behavioural opportunities, and providing more stimulating and responsive environments. Examples range from naturalistic foraging tasks, such as the ubiquitous artificial termite mound, puzzle feeders constructed from PVC pipes, finely chopped and scattered food, novel foods and carcasses, to objects that are introduced for manipulation, play and exploration, novelty and sensory stimulation. Appropriate social stimulation, both within and between species, and even training can be considered as enrichment. On a larger scale, the renovation of an old and sterile concrete exhibit to provide a greater variety of natural substrates and vegetation, or the design of a new exhibit that maximizes the opportunities for natural behaviours, are also considered as environmental enrichment.

(Shepherdson, 2003, p. 119)

From Shepherdson's account above, one may tease out the following theses:

1. It is regarded by its practitioners to be in accordance with the philosophy of promoting animal welfare.[1]
2. It tacitly acknowledges that the zoo environment is very different from the wild environment; we have seen that in the latter, animals have to fend, hunt/forage for themselves while in the former, all services are laid on as far as food, shelter and protection from danger go.

3. It recognises that the animals are bored by the zoo environment, and hence the need to provide for opportunities to stimulate the bored animals.[2]
4. It implies that it is Janus-faced, that is to say, it looks back to the animals's 'behavioural biology and natural history' as well as to the present and the future with regard to their behaviour in their new exotic environment.

In other words, it is at the very centre of the procedures and processes of transforming the inmates to becoming immurated animals, to becoming domesticants. While recognising that the animals carry with them a biology and a natural history peculiar to the species which have evolved naturally in the wild, it also recognises that zoos cannot satisfy the needs which arise from that biology and that natural history in the same way as their ancestors or their wild counterparts can satisfy them in the wild. It cannot do so for two reasons: (a) the simple contingent one that the zoo environment is utterly different from their natural habitat; and (b) the theoretical and logical one that zoos cannot meet the needs of the animals in the way their ancestors meet theirs without abolishing zoos themselves, and returning the animals to the wild. Hence zoos offer naturalistic environments, naturalistic foraging and hunting instead.

As this book has demonstrated, zoo experts and managers hold that zoo animals are 'wild animals in captivity', but such a claim is conceptually speaking an oxymoron, and ontologically speaking wrong-headed and totally misleading because wild animals are naturally occurring beings, whereas zoo animals are biotic artefacts, the ontological foil to the latter. These deep-seated flaws are reflected in the programme and techniques of environmental enrichment which, as we have just remarked, are Janus-faced. In turn, these techniques for ameliorating boredom and thereby improving the welfare of the animals serve to hasten the procedures and processes of transforming them into biotic artefacts, to being domesticants. Over time, as the animals adapt to and evolve within the context of selection for captivity, zoo experts, too, would evolve their techniques of enrichment such that they will increasingly be less and less naturalistic and more and more oriented to human culture rather than to animal culture as found in nature amongst animals living in the wild.

According to Shepherdson (2001) and others, an enriched environment should allow the animals to perform 'natural behaviours' implying that the behaviour of animals in zoos are 'unnatural' or 'abnormal'. But what counts as 'unnatural or abnormal behaviours'? In the widest sense of 'abnormal' or 'unnatural', all behaviours of zoo animals may be said to be abnormal or unnatural – eating zoo pellets as much as playing with plastic toys are abnormal or unnatural, as such behaviours are not found amongst animals-in-the-wild. The kinds of 'abnormal/unnatural behaviour' which zoo experts are keen to eliminate are those which appear to indicate that the animals under their charge are unhappy, such as stereotypic behaviours. That is why one clear goal of environmental enrichment is to improve their psychological well-being as well as physical welfare.[3] Take the case of chimpanzees and the oft-quoted termite mound. In the wild, chimpanzees have been observed to fish termites out of a termite mound with a stick. So, zoos provide the chimpanzees with an artificial termite mound, not, however, containing termites which the chimpanzees can fish out with a stick, but pie fillings, mustard and other Macgoodies instead. This enrichment, undoubtedly, serves to amuse and distract the chimpanzees, thus preventing them from performing stereotypic behaviours out of boredom. It does not, however, instantiate the natural behaviour of chimpanzees in the wild; these actually use a stick to fish out live termites from the termite mound which they have found for themselves, not out of boredom, but as part of their search for food in order to survive. To say the least, pie fillings and mustard are not part of the diet of chimpanzees in the wild. The mound is artificial, without the smell, the texture of a real termite mound in the wild; the Macfoods are mere snacks for which the chimpanzees have, unfortunately, developed a taste. Macfoods are known not to do humans any good; it is also unlikely to do good to the captive chimpanzees, whereas the termites, which chimpanzees in the wild adore, are of significant nutritional value to them. The two contexts are so fundamentally and utterly different from each other that it would be grossly misleading and grotesque to say that fishing for Macfoods from an artificial termite mound in a naturalistic enclosure with human visitors gawping at them is an instance of 'natural behaviour' as performed by chimpanzees in the wild.

It would be better explicitly to admit the ontological gulf between the two contexts and simply adopt whatever techniques zoo managers and experts can devise in order to keep boredom at bay as far as their charges are concerned. So why not, then, opt not for 'natural behaviours' and naturalistic analogues but simply for the most effective way(s), which in certain contexts can be the most charged with human culture? Let's go back to the chimpanzees. Why not allow them to play with computers, even to play with specially devised computer games, if these can be taught to them? Immurated chimpanzees, after all, are part of human culture; it would be in keeping with that logic to let them be exposed to the latest developments in hi-tech should such exposure achieve at least two of the three desired goals of the enrichment programme as outlined by Shepherdson (2001), namely to improve the psychological (and physical) well-being of zoo animals as well as to render them more interesting to zoo visitors.[4] The same goes for elephants. Elephants in the wild do no painting. But elephants in zoos, who are bored out of their minds, should be given access to paint pots and a large canvass, so that they can have a whale of a time tossing the pots and the paint at the canvass with their trunks, thereby creating at the same time works of elephant 'art', which the zoos can auction to the highest bidder – surely, a win-win situation for all the parties concerned, the elephants, the visitors and the owners of the elephants. Of course, it is true that in tossing the pots the elephants are doing something 'natural', so it counts to that very limited extent as 'natural behaviour', since it is with the same trunk and the same action of tossing that the elephants use to toss the pots as their wild counterparts would to toss an offending human against a tree trunk in the wild. But again, the two contexts are so different in ontological character that it would be highly misleading simply to say that the one is as 'natural' a behaviour as the other.

The logic of the above perspective leads to a conclusion which would make zoos more like circuses. Of course, circuses are often condemned for the cruelty of the methods used in training animals for their acts; reputable zoos, undoubtedly, would be against cruelty in any of the methods and devices they use in their enrichment programmes. Furthermore, it offends our human sensibility to see animals in circus acts dressed up in human clothes; reputable zoos, today, do not impose human accoutrements on their animals. However, these important differences apart, the similarities between them remain striking. The performing whales, each indifferently called Shamu, are a case in point. If one must keep whales in aquaria, then one, at least, ought to make sure that they are not bored out of their minds; what better means of making their lives more meaningful than to train them to enter into a relationship with their keepers/trainers who, seemingly without cruelty, manage to get them to perform according to their orders, by making use of movements they would 'naturally' use if they were living wild lives in the wild? Looked at from such an angle, and bearing in mind that these animals are not 'wild animals in captivity' but immurated animals which are on the way to becoming domesticants, there is nothing odd or uncomfortable about their trained behaviour. Their trained behaviour is only odd and uncomfortable to behold if one mistakenly holds them to be wild animals. Similarly, the orang-utans who have been trained to slide down the vine to greet the human guests gathered in their enclosure for breakfast are not all that different from cats and dogs which go into the kitchen to greet their owners when they saunter down for their breakfast. The orang-utans do not appear to be unhappy doing their act; on the contrary, the act is part of the enrichment programme designed to make life less boring for them. Zoos may come to displace circuses as the more acceptable face of training animals to perform acts which appear both to amuse the animals as well as their human spectators. Now, of course, circus trainers have long maintained that their training also plays such dual roles; however, circuses carry a burdensome past which makes it difficult for them to convince modern sensibilities that their methods are those that the philosophy of animal welfare could endorse with a clear conscience, whereas zoos are in a better position to do this PR job.

The direction towards which the analysis points reflects the shift in terminology from 'environmental enrichment' to 'enrichment'. The former conjures up artificial trees and branches in enclosures from which monkeys could swing; it attempts to provide a naturalistic environment for the animals and thus to lessen their boredom by making it possible for them to swing from branch to branch. The latter conjures up the chimpanzee amusing itself with a computer, which focuses on the animal and attempts to lessen its boredom by providing whatever devices or situations that happen to work. The simpler term looks forward to active training – usually involving a close and intimate relationship between the human trainer and the animal – as a means to improve the psychological well-being of immurated animals, while the more restrictive term appears to focus more on improving the physical welfare as well as the psychological well-being of such animals by making their physical environments more naturalistic.

Conclusion

The concept of environmental enrichment or enrichment tacitly acknowledges that zoos are exotic environments for exotic animals. It is at the centre of those procedures and processes under immuration to transform animals to becoming new kinds of domesticants. The logic of environmental enrichment leads to the logic of enrichment. Under that latter logic, the distinction in theory and principle between circuses and zoos would be difficult to make, as both may claim that they are enriching the existence of the animals under their charge as well as enriching us human spectators who watch the animals performing their trained acts.

Notes

Introduction

1. Ontology is that part of philosophy which deals with being, with different kinds of being in the universe. For instance, God, the devil, angels and demons are supernatural, transcendent beings; Hamlet and Anna Karenina are fictional beings; Michael Jackson and Prince Charles, on the other hand, are flesh and blood individuals whom, in principle, one can meet face to face and whose hands one can shake. The latter are material beings, with space-time co-ordinates, even though they may not be wholly physical beings. In contrast, fictional and supernatural beings are, *ex hypothesi*, non-material, non-physical beings. If someone literally claims to have met Anna Karenina in Russia and kissed her hands, that person runs the risk of being incarcerated in what was once called a lunatic asylum. Of course, some people in history have also seriously claimed that they have seen God, that they have spoken to God, or that God has spoken to them – atheists regard such people to be equally suffering from delusions, but religious believers consider them to be very special individuals whom they call mystics.
2. There is more than one sense of tame. 'Immurated' is a term coined to refer to zoo animals, the detailed meaning of which will be explored, in due course, in the book.
3. Note that the operative phrase here is: 'the suspension of the mechanism of natural selection within the context of natural evolution'. We shall see in later chapters that natural selection does operate in contexts outside of natural evolution, such as in the context of zoo management and domestication. The crucial difference between them is that while the former leads to the emergence of naturally occurring living beings, that is, wild animals in the wild, the latter leads to the emergence of biotic artefacts, the ontological foil to wild animals in the wild.

1 What does the public find in Zoos?

1. Some of the issues raised in this chapter may be found in an earlier article – see Lee, 1997b.
2. The ontological significance of this will be explored in Chapter 4.
3. In contemporary literature in ethics, there are three main types of ethical approaches:

   (a) Consequentialist: The most often invoked of which is utilitarianism. Consequentialism considers an act to be right only if its good consequences turn out overall to outweigh its bad consequences. In other words, the notion of good logically precedes that of right.

(b) Deontological: The Kantian variety has been long dominant in modern Western moral philosophy. Deontology considers an act to be right irrespective of its consequences, emphasising the motive of the act. In other words, the notion of right is central to ethical discourse, and not that of good.

(c) Virtue ethics: The Aristotelian variety is most often invoked; it considers character of the agent to be at the heart of ethics, not so much the agent's acts.

For details, see Baron, Pettit and Slote (1997).

4. The term 'morally considerable' primarily refers to beings which we humans deem to have moral needs or to be the bearer of moral rights in virtue of the fact that they possess certain relevant empirical characteristics, such as sentience or mental life.

5. See also Rachels (1991).

6. The term 'human chauvinism' was coined by Richard Routley (Richard Sylvan) (1973) to draw attention to and critically question the anthropocentrism (human-centredness) deeply embedded in Western moral philosophy. Human chauvinism considers human beings alone to be morally considerable.

7. Of course, there are a few zoos which specialise in exhibiting domesticated animals.

8. According to *The World Zoo Conservation Strategy* (1993): 'It is difficult to estimate the total number of animals and species in zoos (5.1).' It only gives the world total of zoo vertebrates as 1 million animals.

9. I owe this point to Mary Midgley.

10. For an account of classical and molecular genetics, see Lee (2005a).

11. As we shall see in a minute, Darwinian evolution and natural selection must not be understood in terms of what Karl Popper has called 'passive Darwinism', that is to say, as if it implies that 'organisms are the mere playthings of fate, sandwiched as it were between their genetic endowment and an environment over which they have no control' (Rose, 1997, 140).

12. On the themes of time and space in biology, apart from Rose (1997), see also Mayr (1982, 71).

13. On reciprocal causation, see Dickens and Flynn (2001); on its equivalent, non-linear causality, see Lee (1989).

14. In Chapter 9, we critically look at genetic reductionism and show that it is methodologically wrong-headed in the context of *ex situ* conservation of endangered species, a flaw which undermines a key role, if not the key role, assigned to zoos by bodies like the World Zoos Conservtion Strategy (1993) and the European Union Zoos Directive (1999).

15. On ecocentrism, see Rolston (1988). In contrast, the philosophy of animal welfare as well as of animal rights are explicitly biocentric in outlook, that is to say that the focus is on the individual animal which is capable of suffering or which actually suffers pain on the one hand, or which is the subject of a mental life on the other. Such a fundamental difference in theory/philosophy between ecocentrism and biocentrism is bound to have implications for policy-making. For instance, the former may not object to letting animals die of hunger under unusually harsh weather conditions, provided they are not

anthropogenic in origin (not caused by humans), whereas the latter would be in favour of saving those individual suffering animals.

Natural biodiversity should be distinguished from another kind which is artefactual, not natural, in character. (See Lee, 2005a and 2004.) One of the main burdens of this book is to argue that zoos are well placed to create artefactual biodiversity and that they are ontologically misguided in claiming that one of their important aims, if not the most important, is to save extant threatened natural biodiversity through their programme of captive-breeding and *ex situ* conservation – see Chapters 9, 11 and Conclusion.

2   Animals in the wild

1. Note, however, that this historical fact of evolution hides two very different types of phenomena which ought to be distinguished – vertical evolution where there is change but without speciation and evolution which involves speciation. According to E. O. Wilson (1994), Darwin was primarily concerned with the former, not the latter – for instance, a genetic mutation in a population of white moths which happens to bestow survival advantage could end up by being one with predominantly black moths. There has been change but no speciation; you start and end with one species. However, Darwin's account of finches in the Galapagos is an instance of vertical evolution with speciation.
2. This is in contrast to the theory of creationism (popular amongst certain fundamentalist Christians) which holds that there is intelligent design in life forms, present and historic, and that God is that intelligent designer and creator.
3. According to Mayr (1982), this is the kernel of truth behind population thinking in biology – that individual (sexually reproducing) organisms are unique, that there is no 'typical' individual, and that mean values are abstractions. Natural selection works on such unique biological individuals.

   The notion of ecosystems is central to ecological thinking. Very briefly, an ecosystem is that ensemble of biotic and abiotic components which interact in a causally reciprocal manner, of which, of course, the animals form an integral part. The boundaries of ecosystems may not be easy to delineate in all cases, but by and large, there is consensus where the limits may be usefully drawn. One should also bear in mind that ecosystems are dynamic, not static in nature. See, for example, Botkin (1990); Botkin and Keller (1995).
4. Apparently, the age of *Homo sapiens* could then be followed by the age of rodents, as these animals can take refuge underground during a nuclear holocaust; moreover, they are better able to withstand radiation than humans and other mammals, should they be exposed to it.
5. For a full exposition, see Lee (1999).
6. See Maturana and Varela (1980) for the introduction of the term 'autopoeisis'.
7. Note that this sense of autonomy has nothing to do with the Kantian sense of autonomy of the (human) will.
8. However, in Chapter 8, we will argue that the intimate ontological link in an organism between existing 'by itself' and 'for itself' has finally been dramatically ruptured by biotechnology – transgenic organisms exist neither 'for themselves' nor 'by themselves'.

9. For example, a mammalian species on average lasts a million years; this kind of extinction is entirely natural, non-anthropogenic, that is to say it is not caused by humans and their activities.
10. There are obviously other senses of the term 'nature' which will not be considered here. For a thorough and detailed clarification of the different senses, see Lee (1999) or Lee (2005b) for a briefer account. Lee (1999) also argues that the distinction between the natural on the one hand, and the artefactual on the other is fundamental given that there are entities in the universe (or more narrowly construed in our solar system) which are totally independent of humans and others, which are the direct products of human intention, ingenuity and manipulation. However, the distinction is meant, in philosophical terms, not as a dualism (in the Cartesian sense) but as a dyadism. Examples of dualisms are: mind/body, male/female, human/ non-human, where the first term mentioned in each set is considered to be superior or to belong to the master class, while the second term refers to an inferior or slave class. The dyadism – naturally occurring/(human) artefactual – which this book is basically concerned with, has no such hierarchical connotations; the dyadic distinction is simply necessary to an understanding of the history of Earth and of life in general on it on the one hand, and of the role played by *Homo sapiens* on the other. Humans belong to a species which happens to possess a unique kind of consciousness, enabling it to develop not only language but also very powerful technologies for transforming the natural to become the artefactual. In other words, human technology makes it possible for humankind to manipulate nature in order to make it embody human intentions and ends. The dyadism in question therefore has ontological import (but has no hierarchical import in terms of superiority or inferiority). The categories of the naturally occurring on the one hand, and the artefactual on the other are distinctly different, ontologically speaking – the former has existed and (in principle) continues to exist and will eventually go out of existence in the absence of humankind, while the latter exists, continues to exist and will exist only as long as humankind itself exists.
11. Chapter 7 will look at the issue whether animals in zoos could be said to be domesticated animals and, if so, in what sense of that term; Chapter 8 will explore the notion of zoo animals as biotic artefacts.
12. For details on the science, philosophy and technology of genetics, see Lee (2005a).
13. Chapter 8 will look more closely at the notions raised here.
14. See Irwin (2001).
15. Three theses of teleology (of which intrinsic/immanent teleology is one) will be distinguished and characterised in greater detail in Chapter 8.
16. The emperor penguins (*Aptenodytes fosteri*) in Antarctica breed during the Southern winter. After a few weeks of courtship, the female lays an egg and then sets off to the sea to feed herself (travelling up to 50 miles or 80 kms), leaving the male, famished for as long as 65 days in the hostile Antarctic environment, to guard and keep the egg warm, and eventually to hatch it. At the end of that long period, the female returns, 'miraculously' recognises her family and immediately starts to feed the recently hatched chick by regurgitating food from her stomach; whereupon her mate leaves straightaway, on his equally long journey, to replenish and refuel himself at sea. See Rockliffe and Robertson (2004).

17. To make empirical and conceptual sense of this kind of phenomena requires the so-called biological-species concept which may briefly be defined as follows: 'a species is a population whose members are able to interbreed freely under natural conditions' (Wilson, 1994, 36). However, this is not to say that the concept is without difficulties. For instance, it is not applicable to organisms (mainly plants) which reproduce asexually.

18. This refers to the deep themes set out in Chapter 1 of the biology of time and history as well as of the biology and philosophy of reciprocal causation of organisms-in-the-environment.

19. See http://www.polarbearsinternational.org [01/12/04]

20. The evolutionary-species concept is different from the biological-species concept which will be raised again in Chapter 8. In other words, there is no one single meaning/definition of species which can do justice to the notion in all the contexts in which it is invoked. One needs to distinguish, at least, between these two different, though related, understandings of the notion. For a fuller discussion of this and related matters, see Mayr (1982, pp. 256–75, 286–7, 295–6). See also Ereshefsky (1998).

21. The token/type distinction will be raised again in Chapter 8, but in the context also of biotic artefacts, as well as assessing whether a biotic artefact such as a captive animal in a zoo could be a said to be a token of a naturally occurring species in the same way a wild animal in the wild may be said to be a token.

22. The variations are both phenotypic and genotypic. Today, with regard to the latter, scientists, in studying the genomes of various species, can ascertain with precision what constitutes the genetic variations between individuals of the species – in the case of the human genome, they have identified what are called SNPs/snips, that is, 'single nucleotide polymorphism'. A SNP represents a DNA sequence variation amongst individuals of a population. SNPs promise to be a money-spinner as they can be used to identify individuals who could be vulnerable to diseases like cancer.

23. After all, as we have earlier remarked, a peacock and a peahen hardly look alike; nor does a caterpillar look like the adult butterfly.

24. There are in fact three species in the wild today: one Asian (*Elephas meximus*), two African – African savannah (*Loxodonta Africana*) and African forest (*Loxodonta cyclotis*).

3　'Wild animals in captivity': is this an oxymoron?

1. A more recent volume is Kleiman, Allen, Thompson and Lumpkin's (1996) *Wild Animals in Captivity*.

2. Increasingly, professionally accredited zoos make it their official policy to use as exhibits zoo-bred animals and would only sanction in exceptional cases the import of animals freshly caught from the wild. In such reputable zoos, only the last two categories of animals would, presumably, form part of their respective collections of exhibits.

3. Of course, 40 years after the event, should the man still be incarcerated, most people in society would say that he should be released especially when his character might even have changed during those long years of imprisonment and he is no longer that dangerously violent man in prey of women. However, the matter is then one of justice and morals, not of conceptual

sense, provided one changes the tense by referring to him as that man who was once dangerously violent denied his freedom and under captivity. Analogously, one could say intelligibly, using the past tense, that this animal under captivity in a zoo was once a wild animal; however, what one cannot intelligibly say, as the arguments in this chapter show, is that such an animal is a wild animal in captivity. The latter is precisely what zoos want to say.

4. As we shall see, animals under long-term captivity could weigh much more than their counterparts in the wild.

5. From now on, whenever appropriate, the term 'animal-in-the-wild' rather than 'animal in the wild' will be used in order to emphasise that the property of wildness in animals can neither be understood apart nor is it detachable from the animal's existence in the wild, as well as to highlight the conceptual incoherence of the term 'wild animals in captivity'.

6. Here is a brief biological summary of tameness and taming, according to one leading zoologist on the subject of domestication:

> Reduced flight distance in the presence of people is one of the most obvious behavioral changes accompanying the domestication process. ... The degree of tameness attained is heavily influenced by the animal's experience with people. Tameness is facilitated when people become associated with positive reinforcers such as food or pleasurable tactile contact. ... they are stressed less by interactions with people and may experience greater reproductive success and productivity. Researchers selecting for tameness in wild silver foxes ... have found predictable changes in the activity of the serotonergic and catecholamine systems of the brain. ... Overall, tameness is becoming one of the better understood behaviours associated with the domestic phenotype.
>
> (Price, 2002, p. 129)

Price also points out the tameness is the single most important effect of domestication of behaviour:

> the single most important effect of domestication on behaviour is reduced emotional reactivity or responsiveness to fear-evoking stimuli (i.e. environmental change). This characteristic is observed in virtually all populations of domestic animals and pervades a wide variety of behavioral responses to both the social and physical environments (e.g. intraspecific social interactions, reactions to the presence of humans, responses to novel objects and places). Reduced responsiveness to fear-evoking stimuli is seen as an adaptation to living in a biologically 'safe' predator-free environment with: (i) limited opportunities for perceptual and locomotor stimulation; (ii) frequent invasions of personal space, with little opportunity to escape from dominant conspecifics; and (iii) frequent association with humans, who are prone to cull untamed and intractable individuals. Available information supports the hypothesis that individuals less reactive to fear-evoking stimuli experience reduced levels of stress in captivity, greater reproductive success, greater productivity (e.g. growth rate, animal products) and are handled by humans with greater ease. ... It is not surprising that one biological trait can be so important to the domestication process. Consider the importance of emotional reactivity to the fitness of animals living in nature.
>
> (Ibid., p. 180)

7. The dodo, because of its evolutionary history, had not encountered enemies which it had to fear, least of all humans; it did not develop flight reaction, nor escape distance. As a result, it became extinct when humans arrived in their habitat, killing them with ease.

8. Hediger (1968, p. 43) also points out that the animal-in-the-wild is capable of assessing very finely and precisely the situation it finds itself in the context of exercising its flight reaction. For instance, an antelope would not necessarily display such a reaction every time it meets a lion; if it judges that its normal predator has already just dined handsomely, it would nonchalantly ignore its presence.

9. Furthermore, Hediger goes on to note perceptively that '(m)an is moreover the only creature able to free himself from the elementary function of escape. By this self-release, man clearly stands apart from the rest of creation, and, as the arch-enemy, is the focus of all animal escape reactions' (1968, p. 49).

10. In recent zoo literature, the term 'tame'/'taming' seems to have dropped out of usage. Instead, it talks of 'desensitization', which in part, if not completely, refers to the same thing. See the following example:

> The first step in teaching an animal to allow husbandry procedures to be performed consists of desensitizing the animal to human touch . . . Another important aspect . . . involves using desensitization techniques to reduce an animal's fear responses to unfamiliar objects and uncomfortable procedures.
>
> (Kuczai II et al., 1998, p. 319)

11. Obviously, domestication which leads to domesticants such as cats and cows involves more elements than taming. The point made here is simply that taming is an essential first step in ultimately producing domesticants or domesticated animals.

    It may be worth drawing the reader's attention to another related matter. Reindeer and yaks are tame, but they are not domesticated animals or domesticants. Some writers have called such animals 'domesticated animals' but not 'domesticants'. This author chooses to use the terms 'domesticated animals' and 'domesticants' interchangeably and would just simply refer to reindeer and yaks as 'tame' which is the antonym of 'wild' in one sense of 'wild'; it follows that reindeer roaming in the Artic north are wild in other senses of 'wild', which is made clear in this book. In other words, 'domesticated animals/domesticants' mean more than just 'tame'; 'wild' means more than just 'untamed', though for an animal to become tame is an essential first stage – a necessary though not sufficient condition – in its transformation to a domesticated animal; for an animal to become tame is to lose one aspect of being wild, though not all.

12. Philosophers talk about the distinction between surface and depth grammar in the following way:

    A1. Sunday of by the way great.
    A2. Sunday is the Muslim holy day.
    A3. Sunday is the Christian holy day.
    A4. Sunday is large in girth.

    A1 is unintelligible at the level of surface grammar as it is not a properly constructed sentence in English. In contrast, A2, A3 and A4 are intelligible at the level of surface grammar because each is a properly constructed sentence in English. A2, though intelligible, happens to be false (if the sentence were to be uttered) while A3 happens to be true. However, it would be inappropriate to say of A4 that it is either true or false, as it is unintelligible at the level of depth grammar – it just makes no sense whatsoever to apply the attribute 'large in girth' literally to Sunday, when Sunday is the name of a day in the week. (Of course it would make sense if 'Sunday' refers to a child, the son of Joe Blogg.)

4 Decontextualised and recontextualised

1. However, it remains true that the public, by and large, are mainly interested in charismatic animals like the lion, the tiger, or cuddly ones like the panda.
2. At least, that is, one of the three species of elephants found in zoos comes from India, the other two from Africa.

   It is true that, historically, the jaguar's home range extended to some of what we today call the southern states of the USA, such as Arizona.
3. Historically, capturing wild animals from the wild to make them residents of zoos also went hand in hand with breeding such animals in captivity; of course, the earlier historical motive for doing the latter is different from the contemporary one which, as Chapter 9 will critically examine, heroically focuses on the goal of saving endangered animals-in-the-wild from extinction. The zoo venture in both aspects began in the nineteenth century in several European zoos – London, Antwerp, Marseilles, Turin – during a period when European imperial power was at its height, making it possible in the first instance for agents from such countries to capture animals-in-the-wild from various parts of their respective colonies, thereby rendering the animals exotic. This set of dislocations involves two things: acclimatisation or naturalisation as well as domestication, which would enable zoos to produce beasts that could be put to work, or crossed with indigenous ones to produce larger and more vigorous versions. (See Baratay and Hardouin-Fugier, 2002.)

   For a discussion of climate on captive animals, see Price (2002, ch. 16).
4. See Appendix for a brief critical exploration of the notion of environmental enrichment.
5. It would be boring and tiresome here to point out once again the unintelligibility of the phrase 'all zoos exhibit living specimens of wild animal species'. The reader should take it as read that, as far as this book is concerned, phrases such as that or similar about the so-called 'natural behaviour' of zoo animals in zoo environments which may occur as part of quotations from zoo literature, are all objectionable.
6. See Baratay and Hardouin-Fugier (2002).
7. This dramatic, theatrical style of exhibiting animals was pioneered by Hagenbeck:

   The panoramas, for which Hagenbeck had received a patent in 1896, were made up of a series of enclosures, laid out like theatre stages, each one behind and slightly higher than the other and separated by hidden moats. Artificial rockwork and plantings concealed the holding quarters and service ways. ... The obscured moats, dramatic rockscapes, and numerous ponds and lakes created scenes of expanding vistas in the most audacious zoo development to that time. The African panorama was the first to generate the illusion of an open savanna, populated with gazelles, flamingos, storks, cranes, antelopes, zebras, lions, and in the distance, ibexes and wild sheep on rocky outcrops.

(Hancocks, 2001, pp. 66–7)

An earlier but no less colourful way of presenting an animal as an exhibit may be found:

[w]hen (Charles X of France) received a giraffe as a gift from ... the Ottoman viceroy of Egypt, in the summer of 1827, he arranged for her to wear a cape embroidered with the French fleur-de-lis and the Egyptian crescent on her walk from the docks in Marseilles to Paris. ... the giraffe's winter quarters were quite elegant, with parquet flooring and the walls insulated with an 'elegant mosaic' of straw matting: 'truly the boudoir of a little lady,' wrote Geoffroy Saint-Hilaire.

(Ibid., pp. 33–4)

8. At first sight, this charge appears unwarranted. For instance, he has a point when he says:

Observing animals in a zoo is often closer to reality when compared with other media, for example film. In a zoo, a pride of Lions ... rests and sleeps for most of the day, just as it would on the African savannah, whereas in a 50-minute-long television programme the Lions are often shown being active for the majority of the time. Therefore, a visit to the zoo provides a more realistic representation of the daily life of a Lion than an edited film programme.

(Andersen, 2003, p. 77)

However, in a later chapter, it will be argued that such a view is at best superficially correct.

9. Chapter 9 will examine these claims from the viewpoint of their mutual compatibility in the light of the ontological exploration pursued by this and following chapters.

10. This argument will be examined in detail in a later chapter.

11. As a result, most animals spend most of the time, throughout the year and especially all nights in barren cages, where they have to put up with a great deal of noise produced by the clanging of steel panels reverberating through the walls – see Hancocks (2001, p. 141).

The cost of creating Jungle World at Bronx Zoo in 1985 was $9.5 million while that of creating Penguin Encounter at Sea World in San Diego in 1983 was $7 million.

12. One such naturalistic exhibit, commonly acclaimed to be the best of its kind, is Jungle World at Bronx Zoo, New York, which opened in 1985. Below is an account of what it really is:

Its real architecture is not the building but 'the design and construction of space ... found inside, and what is inside is a representation of the rainforest, the mangrove swamp and the scrub forest of Asia. . . .' Obviously, this is not an Asian forest, as Asian forests do not grow in New York. Yet in this world full of trees and dense foliage punctuated by the colour of bright tropical flowers through which birds fly and other animals move, and with the rich smell of vegetation and the sounds of a busy jungle, it is almost impossible to remember that one is in fact within a building. Yet it is a man-made forest not just in the sense that trees and plant have been put in a particular location by man, but in the more profound sense that many of the trees are actually manufactured by man. The huge tree which dominates one of the areas is actually made out of steel tubing over which there is metal cloth which is itself covered by an epoxy resin textured and painted so carefully that most people would never guess that it is fake. But the vines which climb around it are real vines. Some vines however are *not*, and those which are provided for the gibbons to swing on are fibre-glass. The mist which envelops the tree tops is real mist but it is produced not by natural conditions, but by the sort of machines used in commercial citrus groves. The rockwork (except for the small pebbles) is artificial but it is a base on which real peat moss and algae grow. Although it might seem that one is in the midst of an undivided tract of forest this is not so; the rocks help form barriers to separate species which may not mix in these conditions. Here one can see animals which actually do live in Asian forests, but what one does not see is the animals living as they would do in that forest. The sound of the cooing of the forest dove is real but it was recorded in Thailand.

(Mullan and Marvin, 1999, pp. 53–4)

13. Apparently, so cleverly done are some of these simulated/naturalistic habitats that not simply are lay visitors taken in by them, but also field biology students. One particular firm of zoo habitat designers, Jones and Jones, has created so realistically the gorilla exhibit in Woodland Park (Seattle) that photographs of it sent to Dian Fossey have fooled her field biology students into believing that the gorillas are wild and in their natural habitat. The National Geographic has also published, in one of its publications on Africa, a photograph of the patas monkey exhibit in that same zoo, in the mistaken belief that it is a piece of photography of wild things in the wild. (See Hancocks, 2001, p. 139.) Such incidents only show how easily humans can be visually misled about what reality is.

14. This reasoning is based, however, not on direct empirical studies of the matter as the author is not aware that any such work has been done. However, it is not implausible to assume that the animal is intelligent enough, given all its faculties and senses with which evolution has endowed it for survival, to work out that there is a discrepancy between immediate perception and reality.

15. See Worstell (2003, ch. 6).

It is not obvious how habitat simulation can re-create the 'essence of a natural habitat'. But this point will be looked at in what follows in the remainder of this section and also in Chapter 5.

16. For some brief details about the technology behind constructing such enclosures, including biodomes, see, for instance, http://www2.ville.montreal.qc.ca/biodome/e1-intro/ef1_rens.htm [15/02/05].

17. That is, until of late, as the food industry becomes interested in the project of manufacturing tastes and flavours.
18. As Mullan and Marvin point out, it is a pity that zoo professionals, on the whole, have not taken to heart what Hediger has said about the limitations of creating naturalistic exhibits:

> The best guarantee of complete naturalness is assumed to be a faithful copy of a piece of natural scenery. This apparently logical conclusion is based on a false ecological estimate that may have serious results. Even an untouched section of the natural ground, enclosed within six sides (i.e. the closest possible imitation of a section of a biotype) is likely to be unnatural ... Mistakes of this kind, resulting in a pseudo-natural arrangement of space, are due to ignorance of the following elementary fact: a cross-section of nature is not an equivalent part of the whole, but merely a piece which, on being completely isolated, alters its quality. In other words: nature means more than the sum of an infinite number of containers of space (cages) however natural.
>
> (Hediger in *Wild Animals in Captivity* as cited by Mullan and Marvin, 1999, p. 77)

It appears that David Hancocks, too, has missed Hediger's point: 'in the animal exhibit areas there must be one constant and inherent design philosophy: Nature is the norm' (Hancocks, 2001, 145). One should not confuse nature with a simulation of nature. If nature were truly the norm, there would be no simulated habitats; indeed, quite simply, there would be no zoos.

5 Lifestyle dislocation and relocation

1. This term does not sound elegant but has been coined by the author; its sense will soon be made clear in the chapter.
2. Of 24 exhibits of mice and rats in the UK, 17 are at London Zoo. Zoo collections, as we have already seen, are not representative of the animal kingdom; if they were, a quarter of their mammals would be bats and a third would be rodents. See http://www.goodzoos.com/Animals/small.htm [12/11/04].
3. See http://members.aol.com/cattrust/cheetah.htm [12/11/04].
4. See Worstell (2003, ch 1).
5. Although the average male qualifies to be megafauna, the female does not. (Any animal weighing more than 100 kg counts as megafauna.)
6. See Clubb and Mason (2002).
7. For further details, see http://www.nwf.org/wildlife/polarbear/;http://www.seaworld.org/infobooks/PolarBears/home.html [13/11/04].
8. For a more detailed discussion of the biological effects of living in miniaturised space under captivity, see Price (2002, ch 17).
9. See Clubb and Mason (2002, 53).
10. See Clubb and Mason (2003). The study makes similar findings with regard to lions, tigers, cheetahs – animals which roam over large areas in the wild – namely, that these are kept in analogously reduced spaces in zoo enclosures, conditions which impinge on the welfare of such zoo animals.
11. A few very rich eccentrics may choose to have no home of their own except for their permanently retained suites at the Ritz, the George V or the Hilton.

12. From this, one is not entitled to draw the conclusion that the need to roam is entirely a parasitic one, a mere side-effect of the need to look for food. In zoos, where the latter need is superseded by hotelification, it would then follow that the need to roam also becomes redundant. However, a recent study shows, for instance, that polar bears in zoos are distressed not simply because they are not allowed to hunt, but also to roam – see Clubb and Mason (2003); see also http://www.admin.ox.ac.uk/po/031001.shtml [01/10/03].

 The idea that the need to roam in the case of mammals or to fly in the case of birds is at best a parasitic one has been very influential and was first made clear by Hediger, who seemed to think that to argue otherwise is merely to be anthropomorphic. He held that it is not a physiologically necessary activity for birds of prey to fly, ignoring the basic understanding that the origin of such a bird's anatomy and physiology stems from its very ability to fly, and that the very organism has been shaped and has evolved within such a context. Hediger was in turn influenced by German philosophy in the 1930s, especially that of Martin Heidegger who gave courses on animals and animality at the University of Freiburg im Bresgau in 1929–30. Heidegger argued that freedom is impossible for animals roaming in the wild as they are consumed with trying to satisfy the essential biological needs of finding food, water, shelter, etc. On the contrary, they are only truly free in zoos as zoos by laying on 'hotel services' relieve them of the need to satisfy such basic functions; consequently, the zoo as such satisfactorily replaces the animal's home range and territory. Such an influential view did not get challenged until the later appearance of animal ethologists such as Lorenz, Tinbergen and others. (On these points above, see Baratay and Hardouin-Fugier (2002, pp. 262–3).

13. The young does so occasionally, but the adult hardly ever.

14. See http://www.worldwildlife.org/gorillas/ecology.cfm [06/01/05].

15. See Clubb and Mason (2002, ch. 4).

16. See http://www.hlla.com/reference/anafr-cheetahs.html [17/12/04].

17. See http://www.nwf.org/wildlife/polarbear/behavior.cfm [18/12/04]. http://www. seaworld.org/infobooks/PolarBears/pbdiet.html [18/12/04].

18. *See* http://www.denverzoo.org/animalsplants/mammal01.htm [10/01/05].

19. See Clubb and Mason (2002, ch. 4).

20. http://www.denverzoo.org/animalsplants/mammal01.htm [10/01/05].

21. The nutritional equivalence is only approximate – according to Clubb and Mason (2002, ch. 4), the zoo diet for Asian elephants contains more fat than the wild diet.

22. It would be too philosophically exhausting as well as unnecessary in this context to defend this thesis in great detail here and now. Suffice it to say that such an assumption is implied by the contemporary programme of (environment) enrichment which enlightened zoo management goes out of its way to emphasise as part of its philosophy. However, the concept of enrichment will be examined in the Appendix.

23. Note that independent value and instrumental value (for humans) are mutually exclusive although the exclusion could be on a continuum – less of one and more of the other; however, this does not mean that a being which has lost its independent value and acquired instrumental value is not a morally considerable being; for example, from the point of view of its ability to suffer pain.

6 Suspension of natural evolution

1. So impressed have the pharmaceutical sciences and industry been by this observation that even a new branch of pharmacology has been established.
2. Carrico (2001); see Plotkin (2000); see also Kuroda (1997) for more examples at http://www.shc.usp.ac.jp/kuroda/medicinalplants.html [11/01/05].
3. This is utilitarian ethics to which reference has already been made in Chapter 1. Although it is not the only normative system, it is, nevertheless, an exceptionally powerful one since the nineteenth century. As we shall see again in a later chapter, the philosophy of animal welfare rests on this axiom: if it is morally good, and therefore, morally obligatory to ameliorate pain in humans who are sentient, it is equally morally good and obligatory to ameliorate pain in all other sentient beings. Some exponents are even keen to extend the reach beyond domesticated and zoo animals to animals-in-the-wild – the logical conclusion to which this perspective can be pushed is a pain-free world where carnivores have been genetically modified to become herbivores, where the lion would literally lie down beside the lamb. For exposition, see Easterbrook (1996); for a critique, see Lee (1999).

    The imperative to save a life (amongst animals) with the same devotion and resources as any human life in peril may be seen in the following account: In 2001, a keeper at Bristol Zoo hand-reared a baby gorilla (Djengi). His mother died soon after his birth. (It is unclear from the account given whether the baby gorilla was found in the wild or that it was captive-born. The title of the article refers to it as 'a wild animal'; but as the word 'wild' is also used when speaking of zoo animals, its reference in this context is none too clear.) The keeper stayed with the infant in the spare bedroom at the house of the keeper of the primate section. The keeper had to feed Djengi every two to three hours in the night; he slept in the keeper's bed for the first few weeks. At about four months, the animal was transferred to a cage put in the living room. His bottles had to be sterilised till he was about 7 months old. He had to be winded until he got big enough to bring up his own wind. The gorilla, like human infants, wore nappies (as it might otherwise foul up the living room). He was bottle-fed Baby formula till 7 months old; then he was fed Complan at bedtime as well. He also had puréed fruit. The keeper sat and watched TV with him; brightly coloured things apparently caught his attention and he liked to play with the remote control. He was also clothed in jumpers. At 9 months, he left Bristol for Stuttgart Zoo, which has been running an orphanage for gorillas for 20 years where the keepers there would continue to look after him until 4 years old, an age when young gorillas in the wild would have become independent of their mothers. Djengi got the same care and attention as a human infant and lived the life of a human infant while at Bristol Zoo. See Wright (2001).

    Caesarean operations are sometimes performed on zoo animals. Jones (2000) reports that The Jersey Zoological Society and Trust (set up by Gerald Durrell) delivered through a caesarean operation of a lioness some lion cubs; the cubs were then bottle-fed.
4. At Emmen Zoo in northern Netherlands, it has recently been reported that two rhinos are being given sun-bed treatment during the winter months in custom-built 4-metre-long sun-beds. As rhinos get older, their skin gets flakier.

Exposing them, especially in the winter, to infra-red sessions of up to 20 minutes and to shorter bouts of ultraviolet rays would improve their skin–blood circulation, as well as give them vitamins. See *The Guardian* (12/02/05, p. 18).

5. In the context of natural evolution (in the wild), natural selection and natural evolution inextricably go hand in hand. Without the former, the latter would not have come about. However, the mechanism of natural selection, nevertheless, does operate in a context other than natural evolution (in the wild) – it may and does operate even in the context of artificial selection, under human-administered selection, for certain characteristics such as colour:

> man does not necessarily select those individuals with the greatest fitness (for captivity) as breeding stock for his selection programs. That is the role of natural selection. Differential mortality and reproduction, including reproductive failure, among artificially selected populations is one way that natural selection in captivity is manifested. . . . [One] focuses on the various ways that natural selection is expressed, namely, mortality and reproductive failure, and changes in these parameters over generations in captivity. . . . [N]atural selection does not cease once a population of animals is brought into captivity, but rather continues to operate regardless of whether or not artificial selection is applied.
>
> (Price, 2002, p. 51)

We should bear the above in mind in Chapter 7 which looks at the notion of artificial selection and domestication.

6. Non-anthropogenic, for the simple reason that they occurred before the evolution of *Homo sapiens sapiens*.
7. Not all animals would be subject to thirst deprivation; cheetahs, for example, do not need to drink water as they get their liquid from ingesting their prey.
8. For a more detailed exploration of synergistic causation, see Lee (1989).
9. The operative phrase here is: 'so-called counterparts in the wild'. This wording leaves it open at this stage of the discussion whether animals-in-the-wild are the true counterparts of zoo animals, a thesis which will be looked at critically in later chapters.
10. See http://members.aol.com/cattrust/cheetah.htm [11/01/05].
11. See http://www.cotf.edu/ete/modules/mgorilla/mgbiology.html [11/01/05].
12. See http://www.seaworld.org/infobooks/PolarBears/pblongevity.html [05/01/05].
13. It is obvious that in such a culture, euthanasia, even with all the practical safeguards against abuse in place, is morally problematic.
14. One such is Hancocks. He talks of a good zoo as one which gives the visitors 'lessons about life' and at the same time 'provide wild animals with safe and contented lives' (Hancocks, 2001, p. 206). *Ex hypothesi*, animals-in-the-wild can never lead safe lives; only animals in zoos can lead lives free from predation, starvation, disease, etc. It makes no sense to say that animals-in-the-wild in their natural habitats can lead contented lives as the lives they lead, with all their hazards and privations, are the only lives they can ever know and can lead. It is possible to say that some zoo animals lead more contented lives than others; those which do not mutilate themselves, do not eat

their own faeces are clearly happier than those which do. However, they engage in such activities only because they have no choice but to live in zoos.

7 Domestication and immuration

1. For the most comprehensive recent account of domestication in all its biological aspects, see Price (2002).
2. As in this context divine design is not pertinent, the discussion will confine itself in this chapter and in the rest of the book to human design only.
3. For details of these historical developments as well as the philosophy of science and technology that they embody, see Lee (2005a).
4. In the case of biotechnology, it may not be so appropriate to talk of breeding new breeds with or minus certain traits, as its techniques permit scientists to insert directly into the genome of another organism or excise from the genome of an organism a DNA sequence that is said to account for the desirable or undesirable trait in the phenotype, especially in the case of single-gene characteristics. While traditional craft-based technology and even Mendelian technology in the case of animals depends on mating, biotechnology by-passes mating altogether; furthermore, while the first two agricultural revolutions lead to the creation and generation of new breeds by mating individuals belonging to different varieties of the same species, biotechnology transcends species and, indeed, even kingdom barriers. (See Lee, 2005a, for details.)
5. See Lee (1969).
6. Note that this author deliberately uses two different words in the two different contexts – 'processes' in the case of naturally occurring events and 'procedures' in the case of technological interventions. Nature involves processes, but uses no procedures. In this usage, 'procedures' imply design and deliberate structuring which have specific outcomes in the mind of the designer.
7. Clutton-Brock (1999, pp. 144, 148) says that the use of elephants in warfare, circuses and zoos and as beasts of burden has a history of at least 3000 if not more than 4000 years.
8. According to one authority (Bökönyi, 1989), one of the constituents of domestication – morphological changes – would have taken up to 30 generations to manifest themselves, at least during the early periods of domestication. Other experts contend that evidence in modern times shows that such changes can take place within a much shorter generational span. See, for example, Bottema and his observations of greylag geese:

> It is well known that after a few generations of domestication greylag geese (*Anser anser*) become fatter and heavier, losing the power of flight ... Besides, after some time, early maturing and loss of the permanent and monogamous pair-bond occurs. Colour variations such as white, piebald, and buff appear, and feet turn orange whereas they were originally pink. The fact that greylag geese become heavier after a few generations is not a genetic change, but a result of feeding. Next to this process, a selection on weight took place, resulting in various extraordinarily heavy breeds.
>
> (1989, p. 32)

9. The adaptation on the part of an animal to its simulated habitat/environment in part involves processes which are natural, and in part procedures which zoo management deliberately imposes on the animals as part of its new existence and lifestyle. An example of the former would be the use of a stick on the part of the chimpanzee to fish for pie fillings, in this case apple sauce, from an artificial termite mound instead of using it to hook up termites from a real termite mound as its relative in the wild would do. This would be an instance of a natural adaptation on the part of the chimpanzee to its (cultural) zoo environment. An example of natural biological processes at work would be the animal becoming heavier than its wild counterpart as a result of the zoo diet, which itself is an instance of zoo policy and zoo procedures. Another instance of the latter would be the obvious fact that freedom to roam is no longer permitted or that a zoo diet is what the animals would get instead of foraging/hunting for their own food.

   Clutton-Brock has made the point more generally as follows:

   > I believe that domestication is both a cultural and a biological process and that it can only take place when tamed animals are incorporated into the social structure of the human group and become objects of ownership. The morphological changes that are produced in the animal follow after this initial integration. The biological process of domestication may be seen as a form of evolution in which a breeding group of animals has been separated from its wild conspecifics by taming. These animals constitute a founder population that is changed over successive generations by both natural and artificial selection, and is in reproductive isolation.
   >
   > (1989, pp. 7–8)

10. There are numerous competing definitions of the term 'domestication'; for a quick discussion see Bökönyi (1989) and Ducos (1989).

11. Note that the passage cited uses the term 'process'; the preferred term, for this author, would be 'procedure'.

12. The pigeon, for instance, is a 'classic case of the exploitation of a symbiotic tendency, for it is essentially a self-domesticating bird which seeks out human fields and settlements' (Issac, 1970, p. 112).

    On the subject of domestication as a symbiotic relationship, see also Budiansky (1994).

13. One way of combating this criticism is to say that the domesticated animal enjoys benefits which its wild ancestor/counterpart would not enjoy, namely that it would have a more secure food supply, be relatively better protected from certain predators and dangers. This, though, might not have been true in all instances especially during the early history of domestication. However, the important point is surely that humans would not have invested consistently over time so much effort, energy and resources to the enterprise unless they believed that domestication would benefit them greatly, irrespective of whether it would also benefit the animals in any way at all. The aim of the exercise is simply to make the animals serve their purposes and goals.

14. One example is the pigeon; see Isaac (1970, p. 112). Another motive which may not necessarily also be economic is ornamentation – some breeds of

dogs, birds and fish fall into this category. There is today another category, animals specially bred or genetically engineered as laboratory animals destined for scientific experiments.

15. Morphological changes are one kind of phenotypical change underpinned in many instances by changes in genetic patterns.

16. He writes:

> Such rapid appearance of deviating colours in many species (of ducks) cannot be explained by mutation during domestication, but it may be due to recessive factors in the wild population. The colour of wild duck species is generally dominant over other colours. The wild-colour pattern is caused by many genes responsible for the various components or for the distribution of the colours. If a mutant factor is present in a duck in heterozygous form, it will not show up in the appearance of the bird, because of the dominance of the wild-colour factors. In practice, the chances of a duck meeting a partner with the same recessive factor are limited: offspring in which combinations of the factor have occurred, e.g. white in homozygous form, will therefore be very rare. Besides there is strong selective pressure against these white mutants, as predators can see them from a great distance, For the same reason, a dominant white mutant will have little chance of surviving. On the other hand, a recessive factor, if present in heterozygous form, is not visible, cannot be eliminated by selection, and thus survives to produce a colour variant only if the owner meets a partner with the same genetic combination. The white trait, a clear negative property in the wild, can be positively valued in captivity by man. As this is a recessive trait, it will be very easy to develop a pure breeding stock of white ducks.
>
> (Bottema, 1989, p. 41)

17. An example from the distant history of domestication, concerning cattle, reinforces the point:

> In cattle, for example, a foreshortened and widened skull, decrease in the dimensions of eye and ear openings, shortness of backbone, decrease in size – in short, overall infantilism – distinguishes domestic from wild varieties. Some of the changes in the soft parts are reflected in skeletal remains. Muscular development or atrophy and changes in brain volume due to environmental modifications, such as differences in food supplied by man or the specialized physiological performance required of domestic animals, mark the skeleton and lead to the development of characteristic crests or ridges.
>
> (Isaac, 1970, p. 21)

18. However, Price (2002) discusses it at great length, although the definition quoted does not.

19. In Chapter 11, we shall be returning to this issue to assess its validity.

20. There are at least four reasons to account for why zoo animals are increasingly captive-bred and zoo-resident for their entire lives. First, as Chapter 9 will argue in detail, zoos exist primarily to exhibit exotic animals to the public; engaging in *ex situ* conservation is a parasitic activity which sits ill with the zoo's main business. Second, not all the animals exhibited in zoos are endangered. Third,

the cost of *ex situ* conservation is prohibitively high. Fourth, current thinking discourages, if not forbids, the replacement of zoo stock by capturing animals-in-the-wild. As a result, world-class zoos engage in captive breeding as a normal method of replenishment; in their bid to be conservation-minded, they hold that the default position must always be no replenishment of stock from the wild. Any deviation from this fundamental prescription will only be permitted under the most stringent conditions and would only be justified under the aegis of *ex situ* conservation. This does not mean, however, that zoos world-wide do not, as a matter of fact, buy animals captured from the wild to replenish their stock – between rhetoric and practice, there is a gulf in many cases.

21. One aspect of acculturation involves responding to the sounds of certain human commands in appropriate ways. One fairly amusing example of this phenomenon has occurred recently when Paris Zoo donated 19 of its surplus zoo-bred baboons to the zoo at Hythe in Kent. The keepers, to their amusement, discovered that these baby baboons only respond to French sounds/words such as 'dejeuner' and 'bonjour' but not to their English equivalents. As a result, the keepers had to buy a French phrase-book to communicate with these animals, who would, probably, remain 'French speaking' for the rest of their lives. (See Ward, 2005.)

22. Recall that in Chapter 6, we were careful in distinguishing between natural selection in the context of natural evolution in the wild on the one hand, and that of natural selection and artificial selection in captive environments on the other. At this stage of presenting the arguments in favour of zoos being an instance of artificial selection in captive environments, it would be appropriate to remind the reader that natural selection can and does occur in captivity.

23. This point is effectively illustrated by the following quotation from a novel by Alexander Dumas which describes the captive (future short-lived) Napoleon II who spent his childhood at the imperial Schönbrunn in Austria. One day, he managed to escape from his allotted quarters in the château to the park surrounding it. The child commented: 'I am as much captive in my room, only that instead of my prison being twenty paces in diameter, it is 3 leagues in circumference. It is no longer my window which is barred; it is that my horizon has a wall' (*The Mohicans of Paris*). (The passage is loosely translated by this author.)

24. The discussion to come and the issue it is meant to elucidate should not be confused with another different and separate issue, namely that deliberately intended artificial selection under domestication may have results which are inadvertent. For example, by consciously keeping animals in captivity, one inadvertently selects for tameness. Conscious selecting for early spawning in hatchery-raised salmon broodstock produces young, which are larger at traditional release times or can be released earlier in the spring. Conscious selecting for the Rex hair colour in rabbits has inadvertently produced certain metabolic and endocrine disturbances which increase mortality and render the animals susceptible to diseases. On these points, see Price (2002, pp. 43–44, 55).

Furthermore, as already noted earlier, the mechanism of natural selection may not entirely be displaced in captivity. It is relevant to cite Price again:

> In general, natural selection in captivity is most intense during the first few generations following the transition from field to captive environments.

Evolution and adaptation to the captive environment occur rapidly during this time because of the change in direction and intensity of natural selection on so many different traits and the relatively large number of correlated characteristics affected. . . . The degree of adaptation to the captive environment will increase as the frequencies of 'favorable' genes increase in response to selective pressure. An improvement in reproduction (i.e. fitness) over the initial generations in captivity can reflect the climb to a new adaptive peak as individuals become increasingly well-adapted to the captive environment over successive generations . . .

(2002, p. 56)

25. The analysis seems to follow roughly the so-called desire-belief model – see Bratman (1999, pp. 6–15).
26. Absolute certainty is 100 per cent; practical or near certainty would be less than 100 per cent though more than what in lay terms may be called high probability. In real life, where diverse and complex variables are at work, the concept of absolute certainty is not appropriate; instead, one operates with the notion of practical certainty.
27. There are exceptions which are covered by what is called strict liability, under which one could be held liable for even unforeseeable consequences, provided one has done something illegal and such consequences flowed causally from that illegal act.
28. For an account of the criminal law (in England and Wales), see Smith and Hogan (1996).
29. Direct intention to kill is first-degree murder; indirect intention to kill attracts a slightly lower category of crime, manslaughter. In jurisdictions where capital punishment obtains, first-degree murder means death by electric shock, chemical means or traditional hanging, while manslaughter means either life imprisonment (with no reprieve) or a fixed sentence (subject to review). However, in jurisdictions which have abolished capital punishment, the distinction between first-degree murder and manslaughter may be more or less academic as both would attract only imprisonment.
30. If such eventualities of failure were to transpire, and if caught, the defendant would be tried for attempted murder, which would merit a verdict of manslaughter.
31. Bratman (1999, pp. 139–42) argues that so-called indirect or oblique intention does not count as intention at all; in his terminology, it would follow that indirect or oblique intention would be said to be 'unintentional'. Bratman's analysis of the concept of intention leads to counter-intuitive results; on his reasoning, the court would have to acquit the woman who posted a kerosene-lit rag through the letter-box of the house of her lover's mistress of either first-degree murder, or indeed, even of manslaughter, as the defendant would, according to Bratman, have acted unintentionally. In the tradition of Anglo-Saxon jurisprudence, at least, no defendant could be found guilty of either first-degree murder or manslaughter if the defendant had acted unintentionally.
32. See Bratman (1999, pp. 143–5).
33. The case of immuration is, therefore, analogous to that of the plane cited earlier. The causal link between direct intention/end (to put animals in captivity

in order to present them as a collection of exhibits to the public on the one hand, to plant a time bomb in the luggage hold in order to claim on the insurance policy on the other) and consequences (bringing about morphological and other biological changes on the one hand, killing the passengers and destroying the plane on the other) is in either case such a strong one that it goes beyond high probability to practical certainty. It is also analogous to the situation in environmental law where the causal link between the polluter's action and his contribution to the pollution amounts to practical certainty.

34. For instance, in ungulates under captivity, emission of adrenaline is low, which in the wild would have hampered the animals from escaping successfully from predators as such low emission reduces the muscular power required for flight. See Baratay and Hardouin-Fugier (2002, p. 273). Genetic decline is also commonly observed in zoos – see Crandall (1964, p. 377); Blomqust (1995, pp. 178–85).

In general, it:

is a reasonably safe assumption that some relaxed selection will accompany the transition from field to captive environments ... Certain behaviors important for survival in nature (e.g. food finding, predator avoidance) lose much of the adaptive significance in captivity. Hence, one would expect natural selection in captivity on such behaviors to lose its intensity. As a result, changes in the gene pool of the population are likely to occur and genetic and phenotypic variability for many traits are likely to increase. For example, behaviors of free-living prey species toward predators may be changed after relatively long periods of freedom from predators ... Caution in accepting novel foods may decline over time in captivity. Free-living herbivores are sometimes exposed to toxic plants ... In contrast, captive animals are generally protected from toxic food items. Hence, it seems reasonable to expect relaxed selection for food neophobia in captive animal populations .... In nature, locating sources of food, water and shelter, mating activities and avoiding predators can require relatively high levels of physical fitness. Physical stamina and agility are less important in captivity due to the absence of predators and provisioning of basic necessities of life by man. ... There is also reason to suspect that natural selection for cognitive abilities may be relaxed in captive populations ... In nature, fitness is enhanced by the ability of individuals to quickly learn the consequences of their behavior or the behavior of other animals. In captivity, humans typically provide animals with the basic necessities for survival and may buffer the negative consequences of their mistakes. Opportunities to exercise cognitive abilities are reduced when the animals' environment limits physical activity and social interactions.

(Price, 2002, pp. 63–5)

35. An omission to do x can be a deliberate act, an act which is directly intended.
36. For the former, see Weilenmann and Isenbrugel (1992).
37. There is no need on the whole for deliberate artificial selection in the breeding of these animals because they are already perfectly well adapted for the task in hand within the environment they are expected to work – see Clutton-Brock (1999, p. 130).

38. Another instance of pertinent evidence concerns the case of hand-reared sloth bears; these 'showed significantly higher frequencies of stereotypic and self-directed behaviours such as masturbation, self-stimulation, and pacing as compared with mother-reared individuals . . .'(Kreger et al., 1998, p. 71) Recently, the newspapers reported a case at the Yangon Zoological Gardens in Burma in which a woman offered to breastfeed two Bengal tiger cubs, four times a day. These cubs had been removed from their mother who had killed the third in her litter. A veterinarian at the Zoological Society of London was reported to have made the following two comments: first, that human milk may lack sufficient fat and protein for a fast-growing tiger cub which can put on as much as 1 kg a day; second, that breastfeeding can cause changes in the animal's behaviour later in life, rendering it a social misfit. See Sample (2005). The latter point is pertinent to our concern here.

8 Biotic artefacts

1. However, this is not to deny that non-human animals make artefacts. We know, for instance, that the beaver makes dams. However, this sort of observation is not germane to the preoccupation of this book which sets out to examine how humans create zoos as an artefact and, in so doing, have also made artefacts of the animals kept and controlled within zoos as exhibits.
2. A more technical and formal definition may be given as follows:

    By an 'artifact' I mean here an object which has been intentionally made or produced for a certain purpose. According to this characterisation, an artifact necessarily has a maker or an author, or several authors, who are responsible for its existence. . . . Artifacts are products of *intentional making*. Human activities produce innumerable new objects which are entirely unintentional (or unintended); such objects and materials are not artifacts in the strict sense of the word. When a person intends to make an object, the content of the intention is not the object itself, but rather some description of an object; the agent intends to make an object of a certain kind or type. Thus what I want to suggest is that artifacts in the strict sense can be distinguished from other products of human activity in the same way as acts are distinguished from other movements of the body; a movement is an action only if it is intentional under some description . . . , and I take an object to be an artifact in the strict sense of the word only if it is intentionally produced by an agent under some description of the object. The intention 'ties' to the object a number of concepts or predicates which define its intended properties. These properties constitute the *intended character* of the object. I shall denote the intended character of an object *o* by '*IC(o)*'.

    Thus an object *o* is a proper artifact only if it satisfies the following *Dependence Condition*:

    . . . The existence and some of the properties of *o* depend on an agent's (or author's) intention to make an object of kind *IC(o)*.

    (Hilpinen, 1995, pp. 138–9)

Note, however, that Hilpinen's definition of 'artefact' is much wider than that used by this book which stipulates that the human intentionality be embodied in a material medium. On this account, unlike Hilpinen's, belief systems and concepts are not artefacts. For instance, the concept *per se* of the division of labour is not an artefact; however, that concept could be applied in practice to design/create, say, the Ford assembly production line which, is, of course, itself an artefact. (For a more thorough philosophical examination of the concept of artefact, see Lee (2005a, ch. 1.)

3. For details of this point, see Lee (2000).

4. Aristotle, in talking about the four causes, has invoked abiotic/exbiotic artefacts to illustrate them; this seems to have influenced unduly how theorists/philosophers have looked at the matter ever since.

5. This view in environmental philosophy is referred to as anthropocentrism, namely that only humans are morally considerable (or intrinsically valuable), and that all other natural things and non-human beings have only instrumental value for humans. We have seen in an earlier chapter that another term for anthropocentrism is human chauvinism.

6. However, no volition should be read into this locution in the case of plants or the lower animals.

7. In this context, *telos* or *tele* (in the plural) is used to refer to the developmental programme, which inheres in every individual organism as a naturally occurring being. For example, an acorn, in accordance with its *telos*, would become an oak sapling, which would grow eventually to be a mature oak tree, producing in turn its own acorns.

8. For degrees of artefacticity, see Lee (1999).

9. However, the term 'species' is not exclusively confined to discourse about biological matters. Historically, it has also been used (although today it appears to have an old-fashioned ring about it) to naturally occurring abiotic matter, such as different natural kinds of minerals. See Wilkerson (1995); see also Laporte (2004).

10. The time-scale is crucial. Those who advocate *ex situ* conservation as a means of saving certain species from extinction are aware of this; hence they talk of a time-span of a 100 years, at most of 200 years, if the captive-bred animals (even in the presence of precautions taken to fend off some of most obvious consequences of immuration) are to remain members of the same species as those individuals which live in the wild.

11. More formally it may be defined as follows: 'a group of actually or potentially interbreeding natural populations that is genetically isolated from other such groups as a result of physiological or behavioural barriers' (Clutton-Brock, 1999, pp. 41–2). Indeed, certain populations of bats may occupy the same space but nevertheless constitute different species, as they do not interbreed because their respective mating calls operate on slightly different frequencies. 'Genetic isolation' is itself a complex notion and may cover numerous aspects of which the bats cited exemplify but one. Another refers to the fact that even if two individuals succeed in mating, it fails to lead to reproduction, and that even should there be successful reproduction, hybrids are born which are sterile, and therefore, in turn, cannot reproduce themselves. For the purpose of this book, the most important aspect is that zoo-born and zoo-bred animals and their counterparts in the wild remain 'genetically isolated'

because they do not and cannot meet and mate, as they occupy different locations/spaces and habitats; they can only meet and mate when humans permit them to do so. (For details, see Lee (1997b).)

12. This author agrees with the view of Mullan and Marvin (1999, p. 12) that zoo animals form a new and distinct species; however, these two authors argue simply on the grounds that zoo animals appear neither to be wild nor domesticated animals. This book argues that they are domesticated, though not in the classical understanding of the term, and that therefore they are immurated animals.

In the literature about classical domestication, variations in the taxonomic designation of the domesticated and the wild are found; for instance, wild and domestic forms of the pig are often given as *Sus scrofa*, on the grounds that although the wild boar and domestic pig are typically found in different environments, their phenotypic differences and their habitat choices are not as dramatically different as between wolves and dogs and that, furthermore, domestic pigs can successfully become feral if given the chance. However, other writers list the domestic pig as *Sus domesticus* – see Clutton-Brock (1999). In general, different taxonomic names commonly mark the distinction between wild and domestic forms. For instance, the domestic chicken is either *Gallus domesticus* or *Gallus gallus domesticus* whereas its wild ancestor, the jungle fowl is called *Gallus gallus*. The controversy concerns the issue whether the domesticated represents a separate species or subspecies. Those arguing for separate status are impressed by the fact that wild and domestic forms are morphologically, behaviourally and/or ecologically distinct. Those arguing against rely on the theoretical possibility of their interbreeding, and therefore on their respective genetic distinctiveness as a subspecies. For further details on this debate and a solution to the problem raised, see Price (2002, pp. 3–4).

9 Justifications deemed serious

1. Vienna was the first to get a (modern) zoo in 1752, followed by Paris in 1793 in the wake of the French Revolution, and then London in 1826.
2. On these points, see Baratay and Hardouin-Fugier (2002).
3. The Department for Environment, Food and Rural Affairs (Defra), UK, has issued a document in connection with the European Union Zoos Directive, 1999. That document contains 7 examples of research projects which zoos could undertake; with the possible exception of one, the rest are concerned solely with issues and problems arising from zoo management and husbandry. See http://www.defra.gov.uk/wildlife-countryside/gwd/zoosforum/handbook/2. pdf, 30–31. [03/02/05].
4. Mayr has written:

> Every organism, whether an individual or a species, is time-bound and space-bound. There is hardly any structure or function in an organism that can be fully understood unless it is studied against this historical background. To find the causes for the existing characteristics, and particularly adaptations, of organisms is the main preoccupation of the evolutionary

biologist. He is impressed by the enormous diversity as well as the pathway by which it has been achieved. He studies the forces that bring about changes in faunas and floras and the steps by which have evolved the miraculous adaptations characteristic of every aspect of the organic world. In evolutionary biology almost all phenomena and processes are explained through inferences based on comparative studies. These in turn, are made possible by very careful and detailed descriptive studies. The evolutionary biologist is interested in the why question.

(Mayr, 1982, p. 71)

5. See Meek (2001).
6. The point made is therefore different to the one made below:

> **Research:** The options available for off-exhibit research animals can actually be more diverse and cost-effective because the emphasis can be purely on functional rather than on aesthetic considerations . . . Public perceptions are not as critical when animals are designated for research purposes and the research facility is not on public view. Environmental enrichment is still important, however, because atypical behaviour and associated physiological stress can add unwanted variation to the experimental design, thereby confounding the results and jeopardizing the validity of the study . . .
>
> (Kreger et al., 1998, pp. 64–5)

While recognising that off-exhibit space for research animals is much smaller and barer than exhibit space and that enrichment is still appropriate, nevertheless these authors have failed to see that the off-exhibit space designated for research animals (which is already smaller than exhibit space) may itself be a cause in bringing about physiological and behavioural changes, and indeed may even lead to brain damage induced by confinement within a very limited space. In other words, enrichment may not be always or entirely successful in counteracting the effects of the variable of severely confined space. More systematic research should be conducted to clarify matters; until that happens, scepticism regarding the validity of results conducted on research animals within confined space, whether in zoos or in laboratories, is justified.

Some zoo researchers and professionals have acknowledged the limitations of enrichment for the psychological well-being of captive animals, such as the macaque monkeys which are the subject of one of these studies:

> Our own studies added substantially to the emerging evidence that modest variations in cage size have little measurable effect on the psychological well-being of monkeys . . . Neither urinary cortisol, appetite suppression, nor abnormal behaviour varied significantly as a function of cage size . . . I want to dispel the illusions that increasing cage size, within the range likely to be possible in a research lab or behind-the-scenes zoo setting, will provide meaningful enrichment to macaques . . . Novelty can stimulate exploratory behaviour but can also elicit fear and disturbance.
>
> (Crockett, 1998, pp. 133–5)

Another set of zoo writers have also admitted the same point made above:

> increasing cage size fails to result in any measurable changes in behaviour . . ., even if that increase is more than 600 times the standard size. The data obtained from these empirical assessments suggest that increasing cage size as a means by which to enrich and enhance an animal habitat may not be worth the cost, at least under conditions in which the size of the cage is the only aspect that is altered.
>
> (Morgan et al., 1998, 160)

7. This seems to be the fate which has overtaken Glasgow Zoo which closed down at the end of September 2003. Admittedly, financial debt was the obvious cause; however, behind that truth is also that the zoo, already in a precarious financial situation, would not be able to meet the requisite demand that it could demonstrate continuous participation in, and contribution to, the goal of conservation.

8. According to the document, all zoos must implement the following measures:

- participating in research from which conservation benefits accrue to the species, and/or training in relevant conservation skills, and/or the exchange of information relating to species conservation and/or where appropriate, captive breeding, repopulation or reintroduction of species into the wild,
- promoting public education and awareness in relation to the conservation of biodiversity, particularly by providing information about the species exhibited and their natural habitats.

9. *In situ* conservation may be defined as:

> The conservation of ecosystems and natural habitats and the maintenance and recovery of viable populations of species in their natural surroundings . . .

Ex situ conservation may be defined as:

> The conservation of components of biological diversity outside their natural habitat.

These definitions are taken from the Convention of Biological Diversity and as cited in Defra's document on the *EU Zoos Directive* (1999, p. 2). http://www.defra.gov.uk/wildlife-countryside/gwd/zoosforum/handbook/2.pdf [03/02/05].

Another similar definition of *ex situ* conservation may be found in *WZCS* (1999). It refers to:

> the maintenance of wild animals in stable populations outside their original biotrope. Being out of their original habitat means that the animals were separated from the other components of their natural community, and are kept in zoos, other types of scientific institutions, breeding centres, or in semi reserves.
>
> (ch. 6.1)

10. Neither do zoos contribute anything significant financially to *in situ* conservation programmes in general, although in 1999 zoos supported more than 650 such projects – see Olney and Fisken (2003).

11. One matter which may be worth pointing out to readers is that the two key bodies (cited in this book) exhorting zoos to embrace conservation as a central justification for their existence differ in their view as to what constitutes breeding stock under *ex situ* conservation. While the *WZCS* stresses that the animals must be captive-born and -bred unless there are exceptional circumstances to justify the capture of an animal from the wild, the Defra gloss on the *EU Zoos Directive*, appears to put the emphasis somewhat differently. It says:

> Stock should only be taken from the wild, regardless of whether it is to be part of a managed programme, if there is evidence to show that collection will not have a detrimental effect on the population, species as a whole or its habitat . . . Collection from the wild is not always detrimental.
>
> (1999, p. 13)

12. A recent telling critique from a zoo professional is that of Hancocks. He says that fewer than five species have been saved from extinction – see (2001, p. xvii). Furthermore, he points out that the 'most optimistic projections state that if all the world's professionally operated zoos, in concert and under perfect conditions, devoted a full half of their facilities to breeding endangered animals they could perhaps manage about eight hundred of them in viable breeding populations' (ibid., p. 152)'. However, of vertebrate species alone, there are about 46,000 in existence.

13. See Article 2 at http://europa.eu.int/eur-lex/pri/en/oj/dat/1999/l_094/l_09419990409en00240026.pdf [03/02/05].

14. As cited by Clubb and Mason (2002, p. 11).

15. Note that captive breeding with the self-contained aim of replenishing zoo stocks is not subject to such a restriction.

16. Of course, domesticants like cats and dogs, the products of what this book has called classical domestication, are another kind of ontological foil to naturally occurring wild animals.

17. As we have seen, which animals get to reproduce is guided by the explicit goal of maintaining genetic variability.

18. There is one outstanding instance of a (private) zoo which runs an *ex situ* conservation programme based on lines which are the exact opposite of what is endorsed by the scientific consensus. This is John Aspinall's Howlett's Animal Park which claims that its success in captive breeding depends exactly on 'being friends' with the animals, encouraging keepers to have intimate physical contact with the animals, romping with them, which makes the animals happy and contented. Now this may be so, as far as the reproductive rate of captive breeding is concerned and as far as animal welfare itself is concerned. However, the point missed by such a perspective is precisely that such intimate contact and relationships with humans render the animals tame – they are happy, contented tame/immurated animals, not wild animals. See http://www.guardian.co.uk/print/0percent2C3858percent2C3962804-103390percent2C00.html (13 February 2000); http://www.totallywild.net/howletts.php?page=howletts. [23/03/05].

19. Just to cite one example regarding the Californian condor (*Gymnogyps californianus*) captive breeding programme initiated by the US Fish and Wildlife Service in 1984, but run by the San Diego Wild Animal Park and Los Angeles Zoo in conjunction with other interested bodies. The 27 last remaining condors (of a reproductive age) were captured to form the core of the breeding programme; from these, the zoos successfully bred more than 200. By 2001, half of them have been returned to the wild and some of these have bred in turn. As mentioned earlier, the chicks were reared by scientists wearing condor-like puppets etc. The first attempt at releasing two of the birds in 1992 was not a success; one of them died when it swallowed antifreeze and the other had to be recaptured when it kept landing on power lines and pylons. The second attempt later that same year was not successful either; three died when they collided with power lines and the other three also had to be recaptured because of their 'fondness' of landing on power lines. The next year, more were released but this time, far from the pylons and electricity lines; however, there was no improvement in the success rate. The scientists finally drew a lesson from these failures – they began to teach their birds to avoid pylons and such dangerous things, by setting up two electricity poles in the enclosures which gave the birds a mild electric shock whenever they approached them. This tactic seemed to have worked as none of the birds in the next batch released died from electrocution or collision with electric cables. But that alone did not ensure long-term survival, as these birds lacked the appropriate knowledge and the relevant skills of how really to survive in the wild, a deprivation brought on by the fact that they were raised by humans disguised as condors within a zoo environment. Eventually, the scientists resorted to giving them in their enclosure what may be called a mentor, an older bird which had been captured from the wild and had known existence in the wild. When these pupils were released, it was found that they behaved more like adult condors. In May 2002, the scientists went even further and released an older bird, which had once lived in the wild and was now well past the age of reproduction, together with the captive-bred juveniles, hoping that she would remember the roosting sites and the watering holes she must have visited 14 years ago. The experiment was acclaimed a success, although problems still remained. The captive-bred birds failed to avoid prey that has been killed with shot which contains lead. Finally in the spring of 2002, the first wild Californian condor was born. By 2020, the scientists expect to achieve the goal of removing the condor from the endangered species list when they will have established two stable wild populations of 150 birds each, as well as maintaining a captive population of 150. See Kaplan (2002).

It is also interesting to note that the reintroduction would not have been possible without the introduction of an exotic related species, namely the Andean condor. It is true that when the programme of reintroducing the Californian condor had stabilised by 1991, the exotics were recaptured and returned to their South American habitat. Nevertheless, this shows that the scientists and related professionals were prepared to import an exotic species in order to learn how condors behave in the wild as well as to enlist these exotics to help the captive-bred juvenile native condors to learn condor culture. (For details, see http://species.fws.gov/species_accounts/bio_cond.html [09/02/05].) This demonstrates the point, which will be discussed in greater detail a little later, that not all conservation scientists fully grasp the true significance of what Chapter 1 of this book has called the zoological conception of an animal, namely that a species in the wild and its individual members, in all aspects of their behaviour and their culture, are the product not only of the processes of natural evolution and the mechanism of natural selection, but also of the complex interrelations among themselves as well as between them and their habitat.

20. For an account of the problems facing captive-born and reared animals for release in nature, see Price (2002), ch. 19, but especially his conclusions on p. 202.

21. For the former, see, for example, http://www.animalinfo.org/species/artiperi/elapdavi.htm [09/02/05]; and for the latter, see Hancocks (2001). As a compromise, one could suggest that the animal became extinct in the wild roughly 1500 years ago.

22. On all these changes, see Price (2002); on genetic changes see also Crandall (1964, 377) and Blomquist (1995, 178–85). Regarding phenotypical changes, Price cites one longitudinal study of wild Norway rats (*R. norvegicus*) over the first 25 generations in captivity. The study reports increase in 'body weight, percentage of mated pairs that produced offspring, number of litters born and length of the reproductive lifespan. The investigators also reported that the tendency to escape and the resistance to handling declined over generations . . .' (Price, 2002, p. 16).

23. See Baratay and Hardouin-Fugier (2002, pp. 273–4).

24. Note that in the case of the Père David deer, no one has a clue not only about its original habitat in the wild but also what the genetic variability in the original wild population would be.

25. An example when conservation scientists/managers forget the equally important precondition of conserving genetic variability concerns the Española tortoise in the Galapagos Islands. From 14 individuals in 1965, the population has increased to over 800 today. Indeed the species would have become extinct by now without the acclaimed success of this captive breeding programme for reintroduction to the wild. However, according to a genetic study conducted by the Free University of Brussels (Belgium), the population lacks genetic diversity. In 1965, the two remaining males and 12 females were transferred to the Charles Darwin Research Station on Santa Cruz Island and captive breeding began with reintroduction in the wild in 1975. However from a study of 134 of this population, the scientist has found that its genetic diversity is equivalent to roughly 11 unrelated individuals when it should have been 300. Nearly 80 of the individuals sampled turn out to have been sired by a single male from San Diego Zoo, nicknamed Super Macho. If this finding is correct, then the revived population may, nevertheless, face extinction in the long run in view of the problems and difficulties facing a population with such a small number of founding members. See Anderson (2004). For a brief summary of why genetic variability is important, see *WZCS* (1999, ch. 6.2).

26. For the term 'lifeline', see Steven Rose, especially ch. 6. It may be appropriate to cite the last paragraph from that chapter:

> Lifelines . . . are not embedded in genes: their existence implies homeodynamics. Their four dimensions [of space and time] are autopoietically

constructed through the interplay of physical forces, the intrinsic chemistry of lipids and proteins, the self-organizing and stabilizing properties of complex metabolic webs, and the specificity of genes which permit the plasticity of ontogeny. The organism is both the weaver and the pattern it weaves, the choreographer and the dance that is danced. That is the fundamental message of this chapter, and therefore in many ways of this entire book. And it provides the framework within which I turn now to consider the mechanisms of evolution.

(1997, p. 171)

27. In July 2004, the Natural Science Museum (London) announced what is dubbed its Frozen Ark Project. The aim of the museum, the Zoological Society of London and the University of Nottingham together with other like-minded institutions throughout the world is to freeze DNA and tissue samples of animals facing extinction, so that scientists would be able to continue to study them from the evolutionary point of view, as well as in the hope that, very soon, advanced cloning techniques would enable the re-creation of extinct animals, using surrogates. The Project will begin with animals from zoos, captive breeding and other research programmes. See Sample (2004) and *New Scientist* (31 July 2004, p. 5).

28. The Indian scientists claim that they have had some breakthroughs recently in overcoming the problems associated with the lack of genetic variability inherent in cloning from cells of a small number of animals.

29. See Ramesh (2004) for the details cited of the Indian cloning project of the Asiatic lion and cheetah.

30. One would not like to carp, but it does seem surprising to read in the two quotations just cited from Hancocks (2001) that its author appears to think that zoos display their animals in natural settings rather than simulated naturalistic settings, that is to say that such exhibition enclosures are constructed entirely with the help of technology, as shown earlier in Chapters 4 and 5.

31. See *WZCS* (1999, ch. 31).

32. According to Kreger et al.:

If an animal or group of animals is intended to serve an educational role, then ... a premium is placed on the naturalistic appearance of both the exhibit and the animals it contains.

(1998, p. 62)

According to Jones and Jones in their 1985 Kansas City Master Plan, their message of structured zoo conservation-education programme is as follows:

A new approach to zoo design begins with presentation of animals in such a way that their right to exist is self-evident. The educational message accompanying this presentation should be clear and persuasive. Whole habitats should be exhibited, with rock and soil substrates and vegetation supporting communities of species typical of the environment and logically associated. Visitors should feel they are passing through a natural environment, with a feeling of intense involvement. The point should be made that animals live in habitats, and it is the destruction of those habitats that is the principal cause of wildlife extinction today.

(As cited by Mullan and Marvin, 1999, p. 60)

33. For instance, according to figures of the *International Zoo Yearbook*, in North America, over 100 million people – that is, just under 50 per cent of the population – visit zoos on an annual basis; the figures for Europe and Japan are similarly high. See *WZCS* (1999, ch. 3.2).
34. The figures are cited in Mullan and Marvin (1999, p. 133).
35. See also ibid., p. 136.
36. This is, however, not to deny that seeing a living exotic animal does have unique appeal (as we shall see in the next chapter); it is just not obvious that it has that special transformative power to educate the public which zoo advocates claim it does.
37. For instance, David Hancocks has written:

careful application of landscape-immersion philosophy, with attention to concealed barriers and authentically replicated forms of the natural landscape and use of borrowed vistas and studied sight lines, can all combine to create a memorably evocative experience in which zoo visitors associate wild animals with appropriate wild habitats. It achieves two important goals. Zoo visitors, even if they don't read the interpretive graphics, can learn by associative intuition that certain animals and certain habitats are inextricable. And they can by similar association gain more respect for wildlife.

(2001, pp. 147–8)

38. The resemblance is not absolute; immurated animals over the years, as we have already mentioned, would display morphological/anatomical and other biological features which are somewhat different from their counterparts in the wild. These differences, however, would not be readily detectable to the passing eye of the ordinary visitor at a distance.
39. See Martin Mere: http://mm.eyelook.co.uk/edu/edu.html [16/02/05].
40. There is another method which is used on the pelicans in St James's Park (London) precisely because the Royal Family found evidence of mutilation, caused by pinioning, stressful. This involves removing one or two strips of the extensor tendons on the leading edge of the wing. This means that the bird would not be able to thrust downwards into the wind to fly. However, this method of rendering it captive but without ostensible mutilation is not foolproof; occasionally in a strong gale, it could be lifted off. See Jones (2002, p. 139).
41. In the case of tamed birds, morphological reductionism does not work so well, as already pointed out the visitors could see readily for themselves that the birds do not fly, and therefore do not resemble in one essential way birds in the wild; they might also observe that one of their wings had been mutilated.
42. One could also get to this conclusion, that a certain dissonance is inherent in the zoo experience itself, even without the benefit of having visited zoos. The experiment one is engaged in is, after all, only a thought experiment, and that is all which is needed here.

43. See Mellen et al. (1998, p. 198).
44. See Baer (1998, p. 293).
45. See Clubb and Mason (2002, ch. 5).
46. One could perhaps say in their defence that they do not distort reality quite as blatantly as Carl Hagenbeck's attempts to create theatrical spectacles of his exhibits at Stellingen at the beginning of the nineteenth century. Hagenbeck exhibited them in the open without cages; he aimed to create the illusion that there was no separation between the human visitors and the captive animals, and indeed between the various species of animals on display. He did not hesitate to put predator next to prey; he created invisible (that is, to the human visitors) barriers which the animals could not cross between the groups of animals. Today's arrangements may spring from different motives and the effects aimed at may not be exactly identical to Hagenbeck's, but in spirit they are akin to his. That is to say that both types of arrangements do not portray animals-in-the-wild, as they manifestly claim to do, but to present immurated animals as exhibits.
47. Note that this author differs from Hancocks in the vital matter of the ontological status of zoo animals, namely, that they are not wild but artefactual in character; on the other hand, Hancocks seems to imply that they are wild, but that, unfortunately, zoos have not succeeded or made the right efforts in presenting them in their full wildness.
48. Such children would be those very same children who see their families buy milk in cartons from supermarkets and, as a result, infer that milk comes from supermarket shelves, and that domesticated animals called cows have nothing to do with the provenance of milk.
49. See Hancocks (2001, p. 249).
50. See Andersen (2003, p. 79).
51. In order to give visitors a taste of carnivores hunting their prey, the Panaewa Zoo in Hawaii has installed clay figures in the tiger enclosure to be operated by a computer but which the public can manipulate, whenever they like, in order to see the tiger 'hunt' a rabbit or squirrel (in clay) which could then be 'saved' by a computer – see Baratay and Hargouin-Fugier (2002, p. 268). Some zoos in non-Western parts of the world are apparently 'bolder' in their policy of what may be fed to their animals in public. This author has been told by a recent visitor to Harbin Zoo that visitors could choose a cow, pay for it, and get the zoo staff to feed it to the tiger for all to behold. Now, this supposedly 'barbaric' practice would at least have the decided merit of teaching the public by allowing them to see what carnivores really eat and how they eat in the wild; most assuredly, they would not leave the zoo thinking that wild tigers in the wild dine off zoo pellets! (This author has no means of checking whether by now such a practice has ceased in Harbin Zoo.)

10 Justifications deemed frivolous

1. Of course, adherents of the philosophy of animal rights would have objections to zoos, but zoo professionals, in the main, are not known to be supporters of animal rights as far as this author knows.
2. These numbers are cited by *WZCS* (1999, ch. 3.2).

3. Ironically, those who may stand a chance of success in bringing a case under the Trade Descriptions Act would be those visitors who can genuinely claim that zoos do not trade fairly only in the light of their zoo experience – that is to say when they fully grasp the dissonance between the exotic exhibits under captivity on the one hand, and wild animals and their behaviour in the wild on the other, and they have been lured by zoos into seeing animals billed as wild when in reality these are not wild at all.

4. Shepherdson writes:

> An animal in the wild basically does only a few things. It hunts, eats, sleeps, often plays and breeds. But when they don't have to hunt for food or engage in normal activities like playing or exploring, they can become bored, even morose. http://www.zooregon.org/ConservationResearch/environm. htm [16/02/05].

5. See Poole (1998, p. 84).

6. Singapore Zoo advertises this experience on the web: http://www.sightseeing-tours.net/web/_li-n73p-2.html [22/02/05]. Tickets start from £13 (Sterling) per person. The highlight of the experience is described as follows: 'Witness the orang utan descend from a naturalistic backdrop filled with vines and branches, followed by a once-in-a-lifetime opportunity for interaction and photography with these intelligent and iconic symbols of the disappearance of the tropical rainforests.' This is followed (without irony) by the line: 'With a commentary highlighting the plight of this magnificent creature, the Zoo hopes to inspire in its guests a respect and deep appreciation of nature.'

7. Possibilities include London Zoo, Singapore Zoo and the Smithsonian National Zoological Park. For the former, see http://www.londonzoo.co.uk/weddings/wedding.html [22/02/05]. Its brochure (received September 2004) says: 'Hold your event at London Zoo and you are helping to support our vital conservation work throughout the world.' Animal Houses: Bear Mountain on the Mappin Pavillion (holds 150 people); Happy Families (200); Lion Terraces (200); Reptile House (350) B.U.G.S! (350); Komodo Dragons (100). 'Unique Features: View some of the world's rarest animals after the zoo closes, in one of these six unique venues. Perfect for canapé receptions, pre-dinner drinks, parties and barbecues.' For Singapore Zoo, see: http://www.singaporebrides.com/venue/wrs/ [22/02/05]. For the Smithsonian Zoological Park, see http://nationalzoo.si.edu/ActivitiesAndEvents/Celebrations/birthday-parties.cfm [25/02/05]. This zoo as well as the San Diego Zoo, amongst others, offer sleepovers for children and adults, marketed as Snore and Roar. http://www. sandiegozoo.org/calendar/cal_sleepovers.html [22/02/05].

8. Shamu is just one of three names – the other two being Kandu and Nandu – dreamt up by the marketing boys to refer indiscriminately to any one of the three performing killer whales in the zoo's entourage, although it is true that Shamu is more often used than the other two. Their real, behind-the-stage, off-duty names are known only to the Sea World staff – see Mullan and Marvin (1999, 23).

11   Philosophy and policy

1. Their philosophical preoccupation focuses on captivity denying the animals the right to be free. Although it is true that not to be free to roam is part of the denial to be wild, nevertheless the source of their concern lies primarily in political philosophy rather than in ontology.
2. Professionally endorsed zoos, at least according to their mission statements, have renounced the policy of capturing wild animals to replenish their stock; instead, they do so through captive breeding.
3. As for the goal of *ex situ* conservation, recall that the last chapter has argued that it is an activity which sits ill with the defining characteristic of a zoo as a collection of animal exhibits open to the public. Therefore, even if such an activity were to be deemed to be desirable, it should be conducted elsewhere in a different space. However, this book has said enough on the subject for the conclusion to be drawn that *ex situ* conservation, because of the ontological risks and the heavy financial resources involved, has little or no merit and that such resources should be channelled towards *in situ* conservation instead.
4. This, however, as an earlier remark has already made clear, should not be interpreted to mean that zoos should not support *in situ* conservation financially or by way of resources which are considered scientifically as well as ontologically relevant to the projects of serving species and their habitats in the wild.
5. It is true that, at the moment, the success rate of such techniques is not very high. However, scientists working in these technologies hope to improve it and are confident that improvement would come in the light of further experimentation and research.
6. The two types of biodiversity – natural and artefactual – have different values because of the difference in their ontological status. See Lee (1999) and Lee (2004).

Appendix: Environmental enrichment or enrichment

1. Shepherdson (2001) gives two other justifications, namely, enriched environments are more interesting to zoo visitors; they help conservation by improving reproductive rates, psychological behaviours when the captive-born grow up to become adults, and survival rates upon introduction to the wild in the case of animals taking part in *ex situ* conservation programmes.
2. However, one should be wary of the claim that boredom is at the bottom of all stereotypies. There may be a genetic 'predisposition for the development of stereotypies' in some cases – see Price (2002, p. 220); furthermore, '[t]here is increasing evidence that many stereotyped behaviours reflect feeding . . . problems . . . that premature weaning and low weaning weight result in the development of relatively high levels of stereotyped wire-gnawing on the lids of their cages' in the case of laboratory mice (*m. musculus*).' In other words, '[s]tress early in life could predispose animals to stereotypy by affecting the persistence of behaviours exhibited at that time' (Price, 2002, pp. 218–19).
3. Physical distress may not manifest itself as stereotypic behaviour but could be readily ascertained by, say, abnormal condition of the feet in the case of

animals with soft pads when they are made to stand on concrete most of the time.

We are, in this discussion, however, concerned primarily with the amelioration of psychological distress.

4. The third, however, namely to aid conservation, only applies to a tiny portion of zoo animals. Furthermore, that small proportion, which does take part in *ex situ* conservation, would have to be protected from such enrichment schemes, as Chapters 7 and 9 have shown.

References and Select Bibliography

Andersen, L. L. (2003) 'Zoo Education: From Formal School Programmes to Exhibit Design and Interpretation', in *Zoo Challenges: Past, Present and Future* (*International Zoo Yearbook*, Vol. 38) ed. P. J. S. Olney and Fiona A. Fisken (London: The Zoological Society of London).

Anderson, James R. (2004) 'Conservation Plans Are Fatally Flawed'. *New Scientist* (24 January 2004): 13. http://www.newscientist.com/article.ns?id=mg18124312.000 [10/02/05].

Baer, Janet F. (1998) 'A Veterinary Perspective of Potential Risk Factors in Environmental Enrichment', in David Shepherdson, J. Mellen and J. Hutchins (eds), *Second Nature: Environmental Enrichments for Captive Animals* (Washington DC: Smithsonian Institution Press).

Baratay, Eric, and Elisabeth Hardouin-Fugier (2002) *Zoo: A History of Zoological Gardens in the West*, trans. Oliver Welsh (London: Reaktion Books).

Baron, Marcia, Philip Pettit and Michael Slote (1997) *Three Methods in Ethics: A Debate*. (Malden, MA, and Oxford: Blackwell).

Blomquist, L. (1995) 'Three Decades of Snow Leopard in Captivity', in *International Zoo Yearbook*, Vol. 34 (1995).

Bököyni, Sandor. (1989) 'Definitions of Animal Domestication', in J. Clutton-Brock (ed.), *The Walking Larder: Patterns of Domestication, Pastoralism, and Predation* (London: Unwin Hyman).

Botkin, David (1990) *Discordant Harmonies: A New Ecology for the Twenty-first Century*. (Oxford and New York: Oxford University Press).

Botkin, David, and Edward Keller (1995) *Environmental Science: Earth as a Living Planet*. (New York and Chichester: Wiley).

Bottema, Sytze (1989) 'Some Observations on Modern Domestication Processes', in *The Walking Larder*, J. Clutton-Brock (ed.) (London: Unwin Hyman).

Bratman, Michael E. (1999) *Intention, Plans and Practical Reason*. (Cambridge, MA.: Harvard University Press).

Budiansky, S. (1994) 'A Special Relationship: The Coevolution of Human Beings and Domesticted Animals'. *Journal of the American Veterinary Medical Society* 204 (1994): 365–8.

California Condor: http://species.fws.gov/species_accounts/bio_cond.html [09/02/05].

Carrico, Christine K. (2001) 'In Search of Pharmacology'. *Molecular Interventions* 1 (2001): 64–5. http://molinterv.aspetjournals.org/cgi/content/full/1/1/64 [10/01/05].

Casamitjana, Jordi (2003) *Enclosure Size in Captive Wild Animals: A Comparison Between UK Zoological Collections and the Wild* (October 2003). http://www.captiveanimals.org/zoos/enclosures.pdf [05/01/05].

Cheetah. http://members.aol.com/cattrust/cheetah.htm. [11/01/05].

——. http://nationalzoo.si.edu/Animals/AfricanSavanna/fact-cheetah.cfm [05/01/05].

——. http://www.hlla.com/reference/anafr-cheetahs.html [06/01/05].

Clubb, Ros, and Georgina Mason (2002) *A Review of the Welfare of Zoo Elephants in Europe* (London: RSPCA). See http://users.ox.ac.uk/~abrg/elephants.pdf [06/01/05].

——. (2003) 'Captivity Effects on Wide-ranging Carnivores', *Nature* 425 (2 October 2003): 473–4. See http://www.nature.com/cgi-taf/DynaPage.taf?file=/nature/journal/ v425/n6957/full/425473a_ fs.html&content_filetype=PDF [06/01/05].

Clutton-Brock, J. (ed.) (1989) *The Walking Larder: Patterns of Domestication, Pastoralism, and Predation.* (London: Unwin Hymaan).

——. (1999) *A Natural History of Domesticated Mammals.* (Cambridge: Cambridge University Press).

Crandall, Lee S. (1964) *The Management of Wild Animals in Captivity.* (Chicago: Chicago University Press).

Crockett, Carolyn M. (1998) 'Psychological Well-being of Captive Nonhuman Primates: Lessons from Laboratory Studies', in *Second Nature: Environmental Enrichments for Captive Animals,* D. Shepherd, J. Mellen and J. Hutchins (eds) (Washington DC: Smithsonian Institution Press).

Department of Farming and Rural Affairs (Defra), UK. http://www.defra.gov.uk/ wildlife-countryside/gwd/zoosforum/handbook/2.pdf [03/02/05].

Denver Zoo. http://www.denverzoo.org/animalsplants/mammal01.htm [10/01/05].

Dickens, W., and J. Flynn (2001) 'Nature or Nurture', *New Scientist* (21 April 2001): 44.

Dipert, Randall R. (1993) *Artifacts, Artworks and Agency.* (Philadelphia: Temple University Press).

Ducos, Pierre. (1989) 'Defining Domestication: A Clarification' in *The Walking Larder,* J. Clutton-Brock (ed.) (London: Unwin Hyman).

Easterbrook, Gregg (1996) *A Moment on the Earth: The Coming Age of Environmental Optimism* (London: Penguin).

Ereshefsky, Marc (1998) 'Eliminative Pluralism', in *Philosophy of Biology,* David Hull and Michael Ruse (eds) (Oxford and New York: Oxford University Press).

Good Zoos. http://www.goodzoos.com/Animals/small.htm [last modified 16 January 2000].

Gorilla. http://www.worldwildlife.org/gorillas/ecology.cfm [06/01/05].

——. http://www.cotf.edu/ete/modules/mgorilla/mgbiology.html [11/01/05].

Hagenbeck, Carl (1910) *Beasts and Men* (London: Longmans and Green).

Hancocks, David (2001) *A Different Nature: The Paradoxical World of Zoos and Their Uncertain Future* (Berkeley, Los Angeles and London: University of California).

Hediger, Heinrich (1950) *Wild Animals in Captivity: An Outline of the Biology of Zoological Gardens,* trans. G. Sircom (London: Butterworths Scientific Publications).

——. (1968) *The Psychology and Behaviour of Animals in Zoos and Circuses,* trans. G. Sircom (New York: Dover Publications).

——. (1970) *Man and Animal in the Zoo* (London: Routledge and Kegan Paul).

Hilpinen, Rosto (1995) 'Belief Systems as Artifacts'. *Monist,* 28 (1995): 136–55.

Hull, David, and Michael Ruse (eds) (1998) *The Philosophy of Biology* (Oxford and New York: Oxford University Press).

Irwin, Aisling (2001) 'Wild at Heart'. *New Scientist* (3 March 2001): 28–31.

Isaac, E (1970) *Geography of Domestication* (Englewood Cliffs, NJ: Prentice Hall).

Jablonka, Eva, and Marion Lamb (2005) *Evolution in Four Dimensions: Genetic, Epigenetic, Behavioral and Symbolic Variation in the History of Life* (Cambridge, MA: MIT Press).

Jones, Oliver Graham (2002) *Zoo Tails* (London: Bantam Books).

Kaplan, Matt (2002) 'The Plight of the Condors'. *New Scientist* (5 October 2002): 34–6.

Kellert, Stephen R., and Joyce K. Berry (1981) *Knowledge, Affection and Basic Attitudes Toward Animals in American Society* (Washington, D.C.: US Government Printing Office).

Kleiman, Devra, Mary E. Allen, Katerina V. Thompson and Susan Lumpkin (eds) (1996) *Wild Animals in Captivity: Principles and Techniques* (Chicago: University of Chicago Press).

Kreger, Michael D., Michael Hutchins and Nina Fascione (1998) 'Context, Ethics, and Environmental Enrichment in Zoos and Aquariums', in *Second Nature: Environmental Enrichments for Captive Animals*, D. Shepherd, J. Mellen and J. Hutchins (eds) (Washington DC: Smithsonian Institution Press).

Kuczai II, Stan A., C. Thad Lacinak and Ted N. Turner (1998) 'Environmental Enrichment for Marine Mammals at Sea World', in *Second Nature: Environmental Enrichments for Captive Animals*, D. Shepherd, J. Mellen and J. Hutchins (eds) (Washington DC: Smithsonian Institution Press).

Kuroda, Suehisa (1997) 'Possible Use of Medicinal Plants by Western Lowland Gorillas (*G. g. gorilla*) and Tschego Chimpanzees (*Pan. t. troglodytes*) in the Ndoki Forest and Pygmy Chimpanzees (*P. paniscus*) in Wamba. http://www.shc.usp.ac.jp/kuroda/medicinalplants.html [11/01/05].

Laporte, Joseph (2004) *Natural Kinds and Conceptual Change* (Cambridge: Cambridge University Press).

Lee, Keekok (1969) 'Popper's Falsifiability and Darwin's Natural Selection', *Philosophy* (1969): 291–302.

——. (1989) *Social Philosophy and Ecological Scarcity* (London: Routledge).

——. (1997a) 'Biodiversity', in *The Encyclopedia of Applied Ethics*, Vol.1, ed. Ruth Chadwick (San Diego: Academic Press), pp. 285–304).

——. (1997b) 'An Animal: What is it?'. *Environmental Values* 6 (1997b): 393–410.

——. (1999) *The Natural and the Artefactual: The Implications of Deep Science and Deep Technology for Environmental Philosophy* (Lanham, MD: Lexington Books/Rowman and Littlefield).

——. (2000) 'The Taj Mahal and the Spider's Web', in *Ethics of the Built Environment*, Warwick Fox (ed.) (London: Routledge).

——. (2004) 'There is Biodiversity and Biodiversity', in *Philosophy and Biodiversity*, ed. Markku Oksanen and Juhani Pieterinen in the series *Philosophy and Biology*, series ed. Michael Ruse (Cambridge: Cambridge University Press).

——. (2005a) *Philosophy and Revolutions in Genetics: Deep Science and Deep Technology*, 2nd edn (Basingstoke: Palgrave Macmillan).

——. (2005b) 'Is Nature Autonomous?', in *Recognizing Nature's Autonomy*, Thomas Heyd (ed.) (New York: Columbia University Press).

MacFarland, D. (1981) *The Oxford Companion to Animals* (Oxford: Oxford University Press).

Markowitz, Hal (1982) *Behavioral Enrichment in the Zoo* (New York: Van Nostrand Reinhold).

Maturana, Humberto, and Francisco Varela (1980) *Autopoiesis and Cognition: The Realization of the Living* (London; Dordrecht: D. Reidel).

Mayr, Ernst. (1982) *The Growth of Biological Thought: Diversity, Evolution and Inheritance.* (Cambridge, MA, and London: The Belknap Press of Harvard University Press).

————. (1988) *Toward A New Philosophy of Biology: Observations of an Evolutionist* (Cambridge, MA, and London: The Belknap Press of Harvard University Press).

Meek, James (2001) 'Cage Life May Drive Lab Animals So Insane that Experiments are Invalid'. *The Guardian* (28 August 2001). http://www.guardian.co.uk/uk_news/story/0,,543234,00.html [03/02/2005].

Mellen, J, and M. S. Macphee (2001) 'Philosophy of Environmental Enrichment: Past, Present and Future'. *Zoo Biology* 20 (2001): 211–26.

Mellen, Jill, Marc P. Hayes, and David J. Shepherdson (1998) 'Captive Environments for Small Felids', in *Second Nature: Environmental Enrichments for Captive Animals*, David Shepherdson, J. Mellen and J. Hutchins (eds) (Washington DC: Smithsonian Institution Press).

Morgan, Kathleen N., Scott W. Line and Hal Markowitz (1998) 'Zoos, Enrichment and the Sceptical Observer', in *Second Nature: Environmental Enrichments for Captive Animals*, D. Shepherd, J. Mellen and J. Hutchins (eds) (Washington DC: Smithsonian Institution Press).

Mullan, Bob, and Garry Marvin (1999) *Zoo Culture*, 2nd edn (Urbana and Chicago: University of Illinois Press).

New Scientist (2004) 'Noah's Freezer' (31 July 2004): 5. http://www.newscientist.com/article.ns?id=mg18324580.700 [10/02/05].

Norton, Bryan et al (eds) (1995) *Ethics on the Ark: Zoos, Animal Welfare and Wildlife Conservation* (Washington and London: Smithsonian Institution Press).

Olney, P. J. S., and Fiona A. Fisken (eds) (2003) *Zoo Challenges: Past, Present and Future* (*International Zoo Yearbook*, Vol. 38) (London: The Zoological Society of London).

Père David deer. http://www.animalinfo.org/species/artiperi/elapdavi.htm [09/02/05].

Plotkin, Mark J. (2000) *Medicine Quest* (New York: Viking).

Polar Bears. http://www.polarbearsinternational.org [01/12/04].

————. http://www.nwf.org/wildlife/polarbear/[05/01/05].

————.http://www.seaworld.org/infobooks/PolarBears/home.html [05/01/05].

Poole, Trevor B. (1998) 'Meeting a Mammal's Psychological Needs', in *Second Nature: Environmental Enrichments for Captive Animals*, David Shepherdson, J. Mellen and J. Hutchins (eds) (Washington DC: Smithsonian Institution Press).

Price, E. O. (1984) 'Behavioural Aspects of Animal Domestication'. *Quarterly Review of Biology* 59 (1984): 1–32.

————. (2002) *Animal Domestication and Behavior* (New York: CABI Publishing).

Rachels, James (1991) *Created From Animals: The Moral Implications of Darwinism.* (Oxford and New York: Oxford University Press).

Ramesh, Randeep (2004) 'Cloning breeds hope for India's big cats'. *The Guardian* (18 August 2004). http://www.guardian.co.uk/international/story/0,,1285280,00.html [09/02/05].

Regan, Tom (1983) *The Case for Animal Rights* (Los Angeles: University of California Press).

Rockcliffe, Wendy, and Graham Robertson (2004) 'Emperor Penguins: Winter Survivors'. http://www.aad.gov.au/default.asp?casid=3524 [01/12/04].

Rolston, Holmes, III (1988) *Environmental Ethics: Duties to and Values in the Natural World* (Philadelphia: Temple University Press).

Rose, Steven P. R. (1997) *Lifelines: Biology, Freedom, Determinism* (London: Penguin).

Routley (Sylvan), Richard (1973) 'Is There a Need for a New, an Environmental Ethic?'. *Proceedings of the XVth World Congress of Philosophy* (1973): i, 205–10.

Ruse, Michael (1988) *Philosophy of Biology Today* (New York: State University of New York Press).

Sample, Ian (2004) 'Frozen Ark to Save Rare Species', in *The Guardian* (27 July 2004). http://www.guardian.co.uk/uk_news/story/0,1269747,00.html [10/02/05].

———. (2005) 'What Is This Woman Doing?'. *The Guardian*, Life Section (7 April 2005).

Shepherdson, D. J. (2005) 'Environmental Enrichment'. http://www.oregonzoo.org/ConservationResearch/environm.htm [16/02/05].

———. (2001) *A Guide to Improving Animal Husbandry Through Environmental Enrichment* (June 2001). http://zcog.org/zcog%20frames/A%20Guide% 20for% 20Improving%20Animal%20Husbandry%20Through% 20Environmental% 20Enrichment/Guia%20Enriquecimiento--Oregon%20Zoo.htm [16/02/05].

———. (2003) 'Environmental Enrichment: Past, Present and Future', in *Zoo Challenges: Past, Present and Future* (*International Zoo Yearbook*, Vol. 38), ed. Olney and Fisken (London: The Zoological Society of London).

Shepherdson, D., J. Mellen and J. Hutchins (eds) (1998) *Second Nature: Environmental Enrichments for Captive Animals* (Washington DC: Smithsonian Institution Press).

Siipi, Helena. (2005) *Naturalness, Unnaturalness, and Artifactuality in Bioethical Argumentation*. (Turku, Finland: Department of Philosophy, University of Turku).

Singer, Peter (1976) *Animal Liberation: Towards an End of Man's Inhumanity to Animals* (London: Jonathan Cape).

Smith, J. C., and Brian Hogan (1996) *The Criminal Law* (London: Butterworth).

Stevens, P. M. C., and E. McAlister (2003) 'Ethics in Zoos', in *Zoo Challenges: Past, Present and Future* (*International Zoo Yearbook*, Vol. 38) ed. P. J. S. Olney and Fiona A. Fisken (London: The Zoological Society of London).

The European Union (EU) Zoos Directive (1999). http://europa.eu.int/eurlex/pri/en/oj/dat/1999/l_094/l_09419990409en00240026.pdf [03/02/05].

The Guardian (2005) 'Sunbed for Rhinos at Dutch Zoo' (12 February 2005): 18.

Tudge, Colin (1992) *Last Animals at the Zoo: How Mass Extinction Can Be Stopped* (Oxford: Oxford University Press).

Ward, David (2005) 'Parlez-vous baboon? Zoo Keepers Bridge Gap'. *The Guardian* (22 January 2005).

Weilenmann, P., and E. Isenbrugel (1992) 'Keeping and Breeding the Asian Elephant at Zurich Zoo', in *The Asian Elephant: Ecology, Biology, Disease, Conservation and Management*, E. G. Silas, Krishnan Nair, M. and G. Nirmalan (eds) (Kerala, India: Kerala Agricultural University).

Weisse, Robert J., and Kevin Willis (2004) 'Calculation of Longevity and Life Expectancy in Captive Elephants'. *Zoo Biology* 23, 4 (August 2004): 365–73. http://www.aza.org/Newsroom/PR_elephantlonglives/ [12/12/04].

Wilkerson, T. E. (1995) *Natural Kinds* (Aldershot: Avebury).

Wilson, E. O. (1994) *The Diversity of Life* (London: Penguin).

The World Zoo Conservation Strategy (1993) sponsored by The World Zoo Organization, The Captive Specialist Group of The World Conservation Union's Species Survival Commission.

Worstell, Carlyn (2003) *Reconciling User Needs in Animal Exhibition Design: Gorilla Exhibits as a Case Study* (ZooLex Zoo Design Organization). http://www.zoolex.org/publication/worstell/gorilla/content.html [revised 14/01/2004].

Wright, Sue (2001) 'What's it like to hand-rear a wild animal?', *Best* (10 July 2001).

Young, Robert (2003) *Environmental Enrichment for Captive Animals*. (Oxford: Blackwell).

Index

Printed in the United States
by Bookmasters

Printed in the United States
By Bookmasters